Chernobyl Prayer

Svetlana Alexievich was born in Ivano-Frankivsk in 1948 and has spent most of her life in the Soviet Union and present-day Belarus, with prolonged periods of exile in Western Europe. She started out as a journalist and developed her own non-fiction genre which brings together a chorus of voices to describe a specific historical moment. Her first book, *The Unwomanly Face of War* (1985), chronicles the experience of Soviet women during the Second World War, while her second volume, *Last Witnesses* (1985), focuses on the same period seen through the eyes of Soviet children. They were followed by *Boys in Zinc* (1991), an account of the effects of war – specifically the Soviet war in Afghanistan – on soldiers, their families and society, and *Chernobyl Prayer* (1997), which features a series of monologues by people who were affected by the Chernobyl disaster. Her most recent book is *Second-Hand Time* (2013), a chronicle of post-Soviet life. She has won numerous international awards, including the 2015 Nobel Prize in Literature for 'her polyphonic writings, a monument to suffering and courage in our time'.

Anna Gunin's recent translations include Oleg Pavlov's award-winning *Requiem for a Soldier* (2015) and Mikail Eldin's war memoirs *The Sky Wept Fire* (2013). Her translations of Pavel Bazhov's folk tales appear in *Russian Magic Tales from Pushkin to Platonov* (2012), shortlisted for the 2014 Rossica Prize.

Arch Tait has translated thirty books, short stories and essays by most of today's leading Russian writers. His translation of Anna Politkovskaya's *Putin's Russia* (2004) was awarded the inaugural PEN Literature in Translation prize in 2010. Most recently, he has translated Mikhail Gorbachev's *The New Russia* (2016).

SVETLANA ALEXIEVICH

Chernobyl Prayer
A Chronicle of the Future

Translated by Anna Gunin and Arch Tait

PENGUIN BOOKS

PENGUIN CLASSICS

UK | USA | Canada | Ireland | Australia
India | New Zealand | South Africa

Penguin Books is part of the Penguin Random House group of companies
whose addresses can be found at global.penguinrandomhouse.com.

First published in Russian as *Чернобыльская молитва: Хроника будущего* 1997
This translation first published in Penguin Classics 2016

018

Text copyright © Svetlana Alexievich, 1997, 2013
Translation copyright © Anna Gunin and Arch Tait, 2016

The moral rights of the author and translators have been asserted

Set in 10.5/13 pt Dante Mt Std
Typeset by Jouve (UK), Milton Keynes
Printed and bound in Great Britain by Clays Ltd, Elcograf S.p.A.

A CIP catalogue record for this book is available from the British Library

ISBN: 978–0–241–27053–0

Contents

We are air: we are not earth

Merab Mamardashvili

Some historical background

Belarus ... To the outside world we remain *terra incognita*: an obscure and uncharted region. 'White Russia' is roughly how the name of our country translates into English. Everybody has heard of Chernobyl, but only in connection with Ukraine and Russia. Our story is still waiting to be told.

Narodnaya Gazeta, 27 April 1996

On 26 April 1986, at 01:23 hours and 58 seconds, a series of blasts brought down Reactor No. 4 of the Chernobyl nuclear power plant, near the Belarusian border. The accident at Chernobyl was the gravest technological catastrophe of the twentieth century.

For the small country of Belarus (population ten million), it was a national disaster, despite the country not having one nuclear power station of its own. Belarus is still an agrarian land, with a predominantly rural population. During the Second World War, the Germans wiped out 619 villages on its territory along with their inhabitants. In the aftermath of Chernobyl, the country lost 485 villages and towns: seventy remain buried forever beneath the earth. During the war, one in four Belarusians was killed; today, one in five lives in the contaminated zone. That adds up to 2.1 million people, of whom 700,000 are children. Radiation is the leading cause of the country's demographic decline. In the worst hit provinces of Gomel and Mogilyov, the mortality rate outstrips the birth rate by 20 per cent.

The Chernobyl disaster released fifty million curies (Ci) of radioactivity into the atmosphere, of which 70 per cent fell upon Belarus. Twenty-three per cent of the country's land became contaminated with levels above 1 Ci/km² of caesium-137. For comparison, 4.8 per cent of Ukraine's territory was affected and 0.5 per cent of Russia's. More than 1.8 million hectares of farmland have contamination levels of 1 Ci/km² or higher; roughly half a million hectares have strontium-90 contamination of 0.3 Ci/km² or above. Two hundred and sixty-four thousand hectares of land have been withdrawn from cultivation. Belarus is a country of forests, but a quarter of its forests and more than half the meadows in the floodplains of the Pripyat, Dnieper and Sozh rivers are located within the radioactive contamination zone.

As a result of constant exposure to low-dose radiation, every year Belarus sees a rise in the incidence of cancer, child mental retardation, neuropsychiatric disorders and genetic mutations.

> *Chernobyl* (Minsk: Belorusskaya
> Entsiklopediya, 1996), pp. 7, 24, 49, 101, 149

Monitoring records show that high levels of background radiation were reported on 29 April 1986 in Poland, Germany, Austria and Romania; on 30 April in Switzerland and northern Italy; on 1 and 2 May in France, Belgium, the Netherlands, Great Britain and northern Greece; and on 3 May in Israel, Kuwait and Turkey.

Gaseous and volatile matter was launched into the sky and dispersed around the globe: its presence was documented on 2 May in Japan; on 4 May in China; on 5 May in India; and on 5 and 6 May in the US and Canada.

In less than a week, Chernobyl became a global problem.

> *Consequences of the Chernobyl Accident in Belarus*
> [*Posledstviya Chernobyl'skoy avarii v Belarusi*]
> (Minsk: International Sakharov Higher College of
> Radioecology, 1992), p. 82

Reactor No. 4, now known as the 'Shelter Object', still holds in its lead-reinforced concrete belly around 200 tonnes of nuclear material. This fuel is partially mixed with graphite and concrete. Nobody knows what is currently occurring inside.

The sarcophagus was hastily constructed to a unique design, one that must fill the design engineers from St Petersburg with pride. It was intended to last for thirty years, but the structure was built with a remote assembly technique, its sections joined together by robots and helicopters: hence the gaps. Today, the data suggest the breaches and cracks exceed 200 square metres in total, and aerosol radioactivity is continually leaking through them. When the wind blows from the north, traces of uranium, plutonium and caesium appear in the south. On a sunny day with the lights off, shafts of light can be seen inside the reactor hall falling from above. What are they? Rain also penetrates the building, and should moisture reach the fuel-containing material, there is the potential for a chain reaction.

The sarcophagus is a corpse which still has breath. It is breathing death. How much longer does it have? Nobody really knows: it remains impossible to access many of the structural components to assess their soundness. What everybody does know is that, should the Shelter Object fail, it would unleash consequences even more devastating than in 1986.

Ogonyok magazine, No. 17, April 1996

From Belarusian online newspaper articles, 2002–2005

Before Chernobyl, the incidence of cancer in Belarus was 82 in 100,000. Today, the rate has risen to 6,000 in 100,000: an almost seventy-four-fold increase.

Over the past ten years, the mortality rate has risen by 23.5 per cent. Just one person in fourteen dies of old age; the majority of deaths occur among able-bodied people in the forty-six to fifty age bracket. In the worst affected provinces, medical screening has shown that for every ten people, seven are in ill health. Travelling through the villages, one is struck by the overspill of the cemeteries.

Many statistics have still not been revealed. Some are so outrageous that they are being kept secret. The Soviet Union sent

800,000 regular conscripts and reservist clean-up workers to the disaster area. The average age of the drafted workers was thirty-three, while the conscripts were fresh out of school.

In Belarus alone, 115,493 people are recorded as clean-up workers. According to figures from the Belarusian Ministry of Health, between 1990 and 2003, 8,553 clean-up workers died. Two people per day.

Here is how the story began.

The year is 1986. The trial of the perpetrators of the Chernobyl disaster is making front-page news in the USSR and abroad.

Now imagine a deserted five-storey block. An apartment building without residents, but full of furniture, clothes and belongings that can never be used again, because this building is in Chernobyl. Yet that block inside the dead city was the venue for a small press conference. It was held by those poised to prosecute the people guilty of the nuclear disaster. The Communist Party's Central Committee had decided at the highest level that the case should be heard at the scene of the crime, in Chernobyl itself. The trial took place in the local House of Culture. Six defendants stood in the dock: the director of the plant, Viktor Bryukhanov; the chief engineer, Nikolai Fomin; the deputy chief engineer, Anatoly Dyatlov; the shift chief, Boris Rogozhkin; the chief of the reactor, Alexander Kovalenko; and the inspector, Yury Laushkin, of the State Nuclear Safety Inspectorate.

The seats for the public were empty: only journalists had come. But then, there were no people left in this closed city: access to the zone of strict radiation control was restricted. Was that why they chose it as the site for the hearing: fewer people meaning less publicity? There were no camera crews or Western correspondents present. Of course, everyone would have liked to see in the dock the dozens of guilty officials, including those in Moscow. Modern science itself should have been called to account. Instead, they settled for scapegoats.

The verdict was delivered: Viktor Bryukhanov, Nikolai Fomin and Anatoly Dyatlov got ten years each. The others received shorter sentences. Anatoly Dyatlov and Yury Laushkin died in detention from the effects of severe radiation exposure. The chief engineer, Nikolai Fomin, went mad. The power plant's director, Viktor Bryukhanov, served his sentence in full, the entire ten years. When released, he was met at the gates by his family and a handful of journalists. The event all but escaped notice.

The former director now lives in Kiev, working as an ordinary clerk for some company.

And that is the end of the story.

Ukraine will shortly start work on a grandiose construction project. A new shelter known as the 'Arch' will be placed over the sarcophagus built in 1986 to cover the destroyed Reactor No. 4 of the Chernobyl nuclear power plant. Twenty-eight donor countries are earmarking an initial tranche of funding for the project of over 768 million dollars. This new shelter is designed to last not thirty, but a hundred years. It is conceived on a grander scale, as it will need to have sufficient capacity for handling waste disposal. The structure will require a massive foundation: concrete pillars and slabs will be used to create what will essentially be an artificial rock base. Next, they will need to prepare the storage that will be used for radioactive waste extracted from beneath the old sarcophagus. The new shelter will be built of high-quality steel capable of withstanding gamma radiation: 18,000 tonnes of metal alone will be needed.

The Arch will be an unprecedented edifice in the history of mankind. The scale of the structure is awe-inspiring: the double shell will tower up to 150 metres. In terms of aesthetics, it will almost be comparable with the Eiffel Tower.

A lone human voice

I don't know what to tell you about. Death or love? Or is it the same thing. What should I tell you about? . . .

We were just married. We'd still hold hands walking down the street, even if we were going to the shops. We were together the whole time. I used to say, 'I love you.' But I couldn't imagine just how much I loved him. I had no idea. We lived in the hostel for the fire station where he worked. On the first floor. Lived there with three other young families. We all shared a kitchen. The fire engines were below us, at ground level. Red fire engines. It was his work. I always knew where he was, what he was up to. In the middle of the night, I heard some noise. There was shouting. I looked out the window. He saw me and said, 'Shut all the windows and go back to bed. The power station's on fire. I won't be long.'

I never saw the explosion itself. Only the flames. Everything was kind of glowing. The whole sky . . . There were these tall flames. Lots of soot, terrible heat. I was waiting and waiting for him. The soot was from burning bitumen, the roof of the power plant was covered in it. He told me later it was like walking on hot tar. They beat back the fire, but it was creeping further, climbing back up. They kicked down the burning graphite. They didn't have their canvas suits on, they left just in the shirts they were wearing. Nobody warned them. They were just called out to an ordinary fire.

It was four o'clock. Five o'clock. Six o'clock. At six, we were planning to visit his parents. We were going to plant potatoes. From our town of Pripyat to Sperizhye, the village where his parents lived, was forty kilometres. He loved planting and tilling. His mum often spoke of how they never wanted him to leave for the town. They even built him a new house. He was called up, served in the Moscow firefighting troops, and when he came back, it was only the fire brigade for him! There was nothing else he wanted to do. (*She falls silent.*)

Sometimes it's like I'm hearing his voice. Like he's alive . . . Even the photos don't get at me the way his voice does. But he's never calling me. Even in the dreams. It's always me calling him.

Seven in the morning. At seven, they told me he was in the hospital. I rushed over, but there was a police cordon round the hospital, they weren't letting anyone in. Only the ambulances were let through. The police were warning us not to go near the ambulances. The Geiger counters were going berserk! I wasn't the only one. All the wives rushed over, everyone whose husband had been at the power plant that night. I ran to look for my friend. She was a doctor at the hospital. I grabbed her by her white coat as she was coming out of an ambulance. 'Let me in there!' 'I can't! He's in a terrible state. They all are.' I wouldn't let go of her: 'I just want to look at him.' 'All right, then,' she says, 'but we'll have to be quick. Just fifteen or twenty minutes.' So I saw him. He was all puffed up and swollen. His eyes were almost hidden. 'He needs milk, lots of it!' my friend told me. 'They need to have at least three litres of milk.' 'But he doesn't drink milk.' 'Well, he will now.' Later, lots of the doctors and nurses in the hospital, and especially the orderlies, came down sick. They died. But back then, nobody knew that would happen.

At ten in the morning, Shishenok, one of the plant's operators, died. He was the first. On that first day. We heard another was trapped under the rubble: Valery Khodemchuk. They never got him out. He was buried in concrete. We didn't know at that time they were only the first.

I said, 'Vasya, what should I do?' 'Get out of here! Just go away! You're having a baby.' I was pregnant. But how could I leave him? He begged me: 'Get out! Save the baby!' 'First I need to bring you some milk, and then we'll see.'

My friend Tanya Kibenok came rushing over. Her husband was in the same ward. She was with her father. He'd brought her by car.

We jumped in and drove to the nearest village to get fresh milk. It was three kilometres out of town. We bought lots of three-litre jars of milk. We got six, so there'd be enough for everyone. But the milk just made them violently sick. They kept losing consciousness all the time. They were put on drips. For some reason, the doctors kept insisting they'd been poisoned by gas, no one said anything about radiation. And the town filled up with these army vehicles, they blocked off all the roads. There were soldiers everywhere. The local trains and the long-distance ones all stopped running. They were washing down the streets with some sort of white powder. I was worried about how I'd get to the village the next day to buy fresh milk. Nobody said anything about radiation. It was just the soldiers who were wearing respirators. People in town were taking bread from the shops, buying loose sweets. There were pastries on open trays. Life was going on as normal. Only they were washing down the streets with that powder.

In the evening, they wouldn't let us into the hospital. There was a whole sea of people. I stood outside his window, he came over and was shouting something to me. Shouting desperately! Somebody in the crowd heard him: they were being moved to Moscow that night. The wives all huddled together. We decided we were going with them. 'Let us see our husbands!' 'You can't keep us out!' We fought and scratched. The soldiers were pushing us back, there were already two rows of them. Then a doctor came out and confirmed they were being flown to Moscow, but he said we needed to bring them clothes – what they were wearing at the power station had all got burned. There were no buses by then, so we ran, all the way across town. We came running back with their bags, but the plane had already gone. They had done it to trick us. So we wouldn't shout and weep.

It was night. On one side of the street were buses, hundreds of them (they were already preparing to evacuate the town), on the other side were hundreds of fire engines. They'd brought them in from everywhere. The whole street was covered in white foam. We were walking over it, cursing and crying.

On the radio, they announced: 'The town is being evacuated for three to five days. Bring warm clothes and tracksuits. You'll be staying in the forests, living in tents.' People even got excited: a trip to the countryside! We'll celebrate May Day there. That'll be something new! They got kebabs ready for the trip, bought bottles of wine. They took their guitars, portable stereos. Everybody loves May Day! The only ones crying were the women whose husbands were ill.

I don't remember how we got there. It was like I woke up only when I saw his mother: 'Mum, Vasya's in Moscow! They took him away in a special plane!' We finished planting the vegetable plot with potatoes and cabbage, and a week later they evacuated the village! Who could have guessed? Who knew back then? That evening, I began throwing up. I was six months pregnant. I was feeling so awful. At night, I dreamed he was calling my name. While he was still alive, I'd hear him in my dreams: 'Lyusya! My Lyusya!' But after he died, he never called my name. Not once. (*She cries.*) I got up in the morning with the idea of going to Moscow on my own. 'Where are you off to in your state?' his mother asked, so upset. They decided his father should get packed too: 'He'll go with you.' They took out all their savings. All their money.

I don't remember the trip. It's gone from my memory. In Moscow, we asked the first policeman we found for the hospital the Chernobyl firemen were in, and he told us. I was quite surprised, because they had been scaring us, saying, 'It's a state secret, top secret.'

It was Hospital No. 6, at Shchukinskaya metro station.

It was this special radiation hospital and you couldn't get in without a permit. I slipped the receptionist some money, and she told me to go on in. She told me which floor. I asked someone else, I was begging them. Then there I was, sitting in the office of Dr Guskova, the head of the radiation department. At the time, I didn't know her name, couldn't hold anything in my mind. All I knew was I had to see him. Had to find him.

First thing she asked me was: 'You poor, poor thing. Have you got children?'

How could I tell her the truth? I realized I needed to hide my pregnancy. Or they wouldn't let me see him! A good thing I was skinny, you couldn't tell by looking at me.

'Yes,' I said.

'How many?'

I thought: 'I've got to say two. If I say one, they still won't let me in.'

'A boy and a girl.'

'As you've got two, you probably won't be having any more babies. Now listen: the central nervous system is severely affected, the bone marrow too.'

'So, all right,' I thought, 'he'll become a bit excitable.'

'And another thing: if you start crying, I'll send you straight out. No hugging or kissing. You mustn't get close. I'll give you half an hour.'

But I knew I wouldn't be leaving this place. I wasn't going anywhere without him. I swore it to myself!

I went in. They were sitting on the bed, playing cards and laughing.

'Vasya!' they called to him.

And he turns round and says, 'Oh no, guys, I'm done for! She's even found me here!'

He looked so funny, had these size forty-eight pyjamas on, though he was a fifty-two. The sleeves and legs were too short. But the swelling had gone down on his face. They were giving them these fluids by a drip.

'What's all this, eh? Why are you done for?' I asked.

He wanted to hug me.

'Sit right back down.' The doctor wouldn't allow him near me. 'No cuddling here.'

Somehow we turned it into a joke. At that point everyone came running over, even from the other wards. All our guys from Pripyat. Twenty-eight of them had been flown here. They wanted to know what was happening back home. I told them they'd begun an evacuation, the whole town was being moved out for three to five days. The guys went quiet. There were two women as well. One had been on reception duty the day of the accident, and she started crying. 'Oh my God! My children are there. What will happen to them?'

I wanted to be alone with him, just for a minute or two. The guys picked up on it, each made some excuse and they went out into the corridor. Then I hugged him and kissed him. He backed away.

'Don't sit near me. Use the chair.'

'Oh, all this is silly,' I told him, waving it off. 'Did you see where the blast was? What happened? You were the first ones there.'

'Most likely sabotage. Somebody must have done it deliberately. That's what all the guys reckon.'

It was what everyone was saying. What they thought at the time.

When I came the next day, they were all in separate rooms. They were strictly forbidden to go out in the corridor or mill about with each other. So they tapped on the walls: dot-dash, dot-dash, dot. The doctors explained that each person's body reacts differently to radiation exposure, and what one person can take would be too much for another. Inside their rooms, even the walls were off the scale. To the left, the right and the floor below they moved everyone out, not one patient stayed. The floors above and below them were empty.

I stayed three days with some friends in Moscow. They told me to take a pot, a bowl, to help myself, not be shy. Amazing people! I made turkey broth for six men. Six of our guys. Firemen on the same shift, the ones on duty that night: Vashchuk, Kibenok, Titenok, Pravik and Tishchura. I picked up toothpaste, toothbrushes and soap for them all at the shops. They had none of that in the hospital. I bought them some little towels. Looking back, I'm amazed at my friends – of course they were frightened, they had to be, what with all the rumours flying, but they still said to take what I needed. They asked how he was doing, how all of them were doing. Would they live? Would they . . . (*She is silent.*) At the time, I met so many good people, can't even remember them all. The whole world shrank to a dot. There was just him. Nothing but him . . . I remember one orderly, she taught me: 'Some illnesses are incurable. You just have to sit and stroke their hands.'

Early in the morning, I'd set out to the market, then back to my friends to boil up some broth. Grated everything, chopped it fine, ladled it out into portions. One man asked me to bring him an apple. I had six half-litre jars to carry to the hospital. Always for six men! Stayed there till the evening. And then back again to the other end of town. How long could I keep it up? But on the fourth day, they told me I could stay in the hotel for medical staff in the hospital grounds. My God, what a blessing!

'But there's no kitchen. How will I cook for them?'

'You won't need to do any more cooking. Their stomachs have started rejecting food.'

He began changing: every day, I found a different person. His burns were coming to the surface. First these little sores showed up inside his mouth and on his tongue and cheeks, then they started growing. The lining of his mouth was peeling off in these white filmy layers. The colour of his face . . . The colour of his body . . . It went blue. Red. Greyish-brown. But it was all his precious, darling body! You can't describe it! There are no words for it! It was too much to take. What saved me was how fast it was all happening, I didn't have time to think or cry.

I loved him so much! I had no idea how badly I loved him! We were just married, couldn't get enough of each other. We'd walk down the street and he'd grab me in his arms and spin me round. And cover me in kisses. The people passing would smile.

He spent fourteen days in the Clinic for Acute Radiation Sickness. It takes fourteen days to die.

On that first day in the hotel, they took readings from me. My clothing, bag, purse, shoes – they were all 'scorching'. So they took the lot away from me on the spot. Even my underwear. They only left me my money. They gave me a hospital gown to put on instead, but it was a size fifty-six – I'm a forty-four – and the slippers were size forty-three, not my thirty-seven. They said I might get my clothing back, or maybe not, they doubted it could be 'cleaned'. So I showed up looking like that. It gave him a fright: 'My God, what's happened to you?'

I came up with a way to cook broth. I put an electric water heater in a glass jar and threw in tiny pieces of chicken. Very finely chopped. Then someone gave me a pot. I think it was the cleaning lady or the hotel attendant. Someone gave a chopping board, which I used for cutting parsley. In that hospital gown I couldn't go to the market, so someone bought me the parsley. But it was all a waste of time, he couldn't even drink. Couldn't swallow a raw egg. And I wanted to give him something tasty! As though it might help him.

I ran to the post office: 'Please, ladies, I have to call my parents in Ivano-Frankovsk urgently. My husband is dying.' Somehow they guessed right away where I was from and who my husband was and they put me straight through. My father, sister and brother flew to Moscow the same day. They brought me my things and some cash.

It was 9 May. He'd always said to me: 'You have no idea how beautiful Moscow is! Specially on Victory Day, when they have the fireworks. I want to show it to you.' I was sitting next to him in the room, he opened his eyes: 'Is it day or night?'

'Nine in the evening.'

'Open the window! The fireworks will be starting!'

I opened the window. We were on the eighth floor, the whole city spread out before us! A bouquet of fire shot into the sky. 'Wow, that's something!'

'I promised that I'd show you Moscow. I promised that I'd always give you flowers for every holiday.'

I turned round and he pulled three carnations from behind his pillow. He'd given some money to a nurse to buy them.

I ran over and kissed him. 'My darling! My true love!'

He grumbled, 'What did the doctors say? No hugging me! No kissing!'

They'd forbidden me to cuddle him or stroke him. But I lifted him and positioned him on the bed. Smoothed the bed sheets for him, took his temperature, brought the bedpan, then took it away. Wiped him down. All night long, I was close by. Watching over every move he made, every sigh.

It's a good job it happened in the corridor and not in his room. I started feeling dizzy and grabbed on to the windowsill. A doctor was walking past, he took hold of my arm. And suddenly he asked: 'You're pregnant?'

'No, no!' I was terrified that someone would hear us.

'Don't pretend,' he said, with a sigh.

I was so shaken that I didn't manage to ask him to keep quiet.

The next day, I was called to the head of the department.

'Why did you trick us?' she asked harshly.

'I had no way out. If I'd told you the truth, you'd have sent me home. It was a little white lie!'

'What on earth have you done!'

'But I'm by his side . . .'

'You poor, poor thing!'

Till my dying day, I'll be grateful to Dr Guskova.

The other wives also came, but they weren't allowed in. The mothers were with me: they let the mothers in. Vladimir Pravik's mother kept begging God, 'Take me instead.'

Dr Gale, this American professor. He did the bone marrow transplant. He comforted me, saying there was hope, maybe not much, but with his strong body, such a hefty guy, we still had a chance. They sent for all his family. Two sisters came from Belarus, and his brother from Leningrad, where he was serving in the army. Natasha, the younger one, was just fourteen, she was crying a lot and frightened. But her bone marrow was the best match. (*She falls silent.*) I'm able to talk about it now. Before, I couldn't. I kept quiet for ten years. Ten years . . . (*She is silent.*)

When he found out the bone marrow would be from his little sister, he flat out refused: 'No, I'd rather die. Leave her alone, she's just a kid.' His older sister, Lyuda, was twenty-eight, she was a nurse and knew what she was going into. 'I just want him to live,' she was saying. I watched the operation. They were lying side by side on the table. The operating theatre had a big window. It lasted two hours. When they'd finished, Lyuda was worse off than him, she had eighteen puncture holes in her chest and had a rough time coming round. And she's in poor shape now, she's registered disabled . . . Used to be this beautiful, strong woman. She never got married. So I was rushing from one ward to the other, from his bedside to hers. By then, he wasn't in an ordinary ward, they'd put him in this special pressure chamber, behind a see-through plastic curtain, which you weren't allowed past. It was specially equipped so you could give injections and insert catheters without having to go behind the plastic. It was all sealed off with locks and velcro, but I worked out how to open them up. I'd quietly move aside the plastic and sneak in to see him. In the end, they just put a little chair for me by his bed. He got so bad that I couldn't leave his side. He kept calling my name: 'Lyusya, where are you? My Lyusya!' He called over and over. The pressure chambers for the rest of our guys were being looked after by soldiers, because the orderlies were refusing and demanding protective clothing. The soldiers took out the bedpans, they mopped the floors, changed the sheets. Took full care of them. Where had these soldiers come from? I didn't ask. I only saw him. Nothing but him . . . And each day I'd hear: 'This one's died, that one's died.' Tishchura died. Titenok died. 'Died . . .' It was like a hammer hitting your head.

He was passing stools maybe twenty-five, thirty times a day. All bloody and gooey. The skin on his arms and legs was cracking. His whole body was coming up in blisters. When he turned his head, clumps of hair were left on the pillow. But he was still my love, my precious one. I tried joking: 'It'll make life easier. You won't need to carry a comb.' Soon they all had their hair cut off. I cut his hair myself. I wanted to do everything myself. If I could have coped physically, I'd have been with him twenty-four hours a day. I felt sorry for every minute away from him. Every minute . . . (*She buries her face in her hands and falls silent.*) My brother arrived and he was scared for me: 'I won't let you go in there!' But Dad says to him: 'You reckon you'll stop her? She'll climb through the window! Up the fire escape!'

I left him, and when I came back, there was an orange on the table. A really big one, pink rather than orange. He smiled. 'Somebody gave it to me. You have it.' The nurse motioned through the curtain that the orange couldn't be eaten. Once it had been lying near him, you couldn't even touch it let alone eat it. 'Go on, eat it,' he said. 'You love oranges.' I took the orange. And just then he closed his eyes and dozed off. They were always giving him injections to sleep, giving him drugs. The nurse looked at me in horror. And me? I was ready to do anything just to stop him thinking about death. Stop him thinking his illness was horrid and that I was scared of him. There's one conversation I remember. Someone was pressuring me: 'You mustn't forget this isn't your husband, it isn't the man you love, it's a highly contaminated radioactive object. You're not a suicide case. Pull yourself together.' But I was like a crazy woman: 'I love him! I love him!' While he was asleep, I whispered, 'I love you!' Walking about the hospital courtyard, 'I love you!' Carrying the bedpan, 'I love you!' I thought back to our life together in the hostel. He could only fall asleep at night holding my hand. It was his habit: he used to hold my hand while he slept. The whole night long.

And in the hospital, I used to hold his hand and wouldn't let go.

It was night. The room was quiet. We were alone. He looked at me really closely, and suddenly he said, 'I want to see our baby so badly. I want to know what he looks like.'

'What should we call him?'

'You'll have to think of something on your own.'

'Why just me, if there are the two of us?'

'Okay then, if it's a boy, let's call him Vasya, and if it's a girl, Natasha.'

'What do you mean, "Vasya"? I already have a Vasya: you! I don't need another.'

I still didn't realize how badly I loved him! It was just him, nothing but him. Like I was blind! Didn't even feel the kicks under my ribs. Though I was already in my sixth month. I thought she was safe and protected inside me. My little one . . .

None of the doctors knew I was staying with him all night in the pressure chamber. They didn't catch on. But the nurses let me. At first, they tried to talk me out of it. 'You're so young. What on earth has got into you? He isn't a person now, he's a nuclear reactor. You'll both frazzle together.' But I followed them around like a puppy. I stood for hours by the door, begged and begged them. And finally they said, 'Oh, to hell with it! You're nuts.' In the morning, just before eight, when the doctors began their rounds, they'd motion through the plastic: 'Quick!' I'd run back to the hotel for an hour. And from nine in the morning till nine at night, I had a permit. My legs went blue right up to the knees. I was so worn out they were swelling up. My soul was tougher than my body. Oh, my love . . .

They didn't do it while I was with him. But when I was gone, they photographed him. He had nothing on, he was naked. Just one thin sheet over him. Each day, I changed the sheet, and by the evening it was all covered in blood. I'd raise him up and bits of his skin would be left sticking to my hands. I asked him, 'Help me, sweetheart! Lean on your hand or elbow, as much as you're able, so I can smooth the sheet, get rid of all the seams and creases.' Just one little seam could injure him. I clipped my nails to the quick so they wouldn't catch on him anywhere. None of the nurses dared go near him or touch him; if they needed anything, they called me. And they took photos of him. Said it was for science. I wanted to kick them all out of there! I wanted to scream and punch them! How could they! If only I could have kept them out. If only.

Coming out of his room into the corridor, I'd almost bump into the wall or the couch, because I couldn't see anything. I'd stop the nurse on duty: 'He's dying.' She'd say, 'What do you expect? He's had 1,600 roentgens, and the lethal dose is 400.' She felt sorry too, but it wasn't the same. For me, this was my love. My sweetheart.

When they'd all died, they refurbished the hospital. Scraped down the walls, ripped up the parquet and got rid of all the woodwork.

I remember only odd snatches from the end. It's all a bit hazy.

I sat through the night at his side on the chair. At eight in the morning, I said, 'Vasya, love, I'm off now. I'll get some rest.' He opened and closed his eyes – that meant he was letting me leave. The moment I used to reach the hotel, make it to my room and flop down on the floor – I couldn't lie on the bed because I was aching all over – an orderly would come banging on the door: 'Quick! Come back! He won't stop calling for you!' But on that morning, Tanya Kibenok was just begging me to go with her: 'Come to the cemetery. I really need you there.' They were burying Viktor

Kibenok and Volodya Pravik that morning. Viktor and he were friends, we were all family friends. Just the day before the explosion, we'd had a picture taken together in the hostel. They looked so handsome in that photo, our husbands! So bright and smiling. It was the last day of our old life. Life before Chernobyl. We were so happy.

I got back from the cemetery and quickly rang up the nurse on duty: 'How is he doing?' 'He died fifteen minutes ago.' No! I'd stayed with him the whole night long. I'd only been gone for three hours! I stood by the window and screamed: 'Why? Why this?' I looked up at the sky and screamed. Screamed the building down . . . People were frightened of coming near. I got a hold of myself and realized: I could see him! One last time! I tore down the stairs . . . He was still lying in the chamber, they hadn't taken him. His last words were 'Lyusya! My Lyusya!' 'She's just slipped out, she'll be back any minute,' the nurse comforted him. He gave a sigh and was gone.

This time there was no dragging me away. I stayed with him right to the grave. Though I can't remember the coffin, just a big plastic sack. That sack . . . In the mortuary, they asked, 'Do you want to see what he'll be wearing?' Yes! They put him in his dress uniform, with the service cap on his chest. They didn't pick any shoes out because his feet were too swollen. He had balloons for legs. They had to slit the dress uniform too, they couldn't pull it on his mess of a body. All just one gory wound. The last two days at the hospital, I'd lift his arm and the bone would be all wobbly, hanging loose, the tissue falling away from it. Pieces of lung, lumps of his liver were coming up through his mouth. He was choking on his own innards. I'd put a bandage on my hand and slip it into his mouth, scoop it all out . . . You can't describe it! There are no words! It was too much to take. This was my sweetheart, my love . . . Not one pair of shoes would fit him. They put him in the coffin barefoot.

Right before my eyes they shoved him in his dress uniform into the plastic sack and tied it up. Then they put the sack in a wooden coffin. And they wound that coffin in another plastic sack. It was transparent plastic, but as thick as oilcloth. Then the whole bundle went into a zinc coffin, they could barely squeeze it in. Just the service cap was left on top.

Everyone came to Moscow, his parents and mine. We bought black headscarves in the city. We were seen by the emergency commission. They told everyone the same thing: that they couldn't give us the bodies of our husbands and sons, they were highly radioactive and would be buried by some special method in a Moscow cemetery. In sealed zinc coffins, under slabs of concrete. And we had to sign some paperwork, they needed our consent. They drummed it into anyone who was unhappy and wanted to take the coffin back home that the dead were now heroes and no longer belonged to their families. They were public property, belonged to the state.

We got into the hearse, us relatives and some army men. There was a colonel, with a walkie-talkie. We could hear someone's voice: 'Await our orders!' For two or three hours, we were driving around Moscow, round the ring road. Then we came back into town. Over the walkie-talkie, we could hear: 'Access to the cemetery is denied. The cemetery is besieged by foreign journalists. Continue to wait.' Our parents kept quiet. My mum was in a black headscarf. I felt like I was about to faint. I had hysterics. 'Why do you have to hide my husband? What is he – a murderer? A criminal? A convict? Who is it we're burying?' My mum said, 'Shush, love.' She stroked my head, held my hand. The colonel radioed: 'Request permission to proceed to the cemetery. The wife is hysterical.' At the cemetery, we were surrounded by soldiers, had to walk under escort. The coffin was carried under escort. No one was allowed to say their farewells, only the relatives. They immediately started filling in the graves. 'At the double!' an officer ordered them. They didn't even let me embrace the coffin. And then we were bundled into buses.

They quickly bought and gave us our tickets home for the next day. The whole time, we had a man in plain clothes with us, carried himself like a soldier, wouldn't even let us leave the hotel room to buy food for the trip. God forbid we should start talking to anyone, especially me. As if I was in a fit state to talk: I couldn't even cry.

When we left, the hotel attendant counted all the towels and sheets, put them in a plastic bag. They must have burned them. We paid for the hotel ourselves – for all fourteen days.

Fourteen days in the radiation sickness clinic. It takes fourteen days to die.

Back home, I fell asleep. I went into the house and collapsed on the bed. I slept for three days straight. They couldn't rouse me. They called an ambulance. 'No, she isn't dead,' the doctor said. 'She'll wake up. It's just a horridly long sleep.'

I was twenty-three.

There's a dream I remember. My dead grandma is there, she's wearing the clothes we buried her in. And she's decorating a New Year tree. 'Granny, why have we got a New Year tree? It's summer.' 'That's just how it is. Your Vasya is joining me soon.' He grew up by the forest. I remember a second dream. Vasya is all in white, and he's calling Natasha's name. Our little girl, who still hadn't been born. She is big already, and I am puzzled at how she's grown so much. He is tossing her up to the ceiling; they are laughing. I am watching them and thinking how simple happiness is. How simple! And later there was another dream. Vasya and I were wading through water. We kept walking on and on. He must have been asking me not to cry. He was sending me a sign from there. Up above. (*She is silent for a long time.*)

Two months later, I travelled to Moscow. Went straight from the station to the cemetery. To see him! And right there in the cemetery, my contractions started. The moment I began speaking to him. They called an ambulance, I told them the address. I gave birth in the same place. In Angelina Guskova's department. She'd said back then: 'When you go into labour, come to us.' And where else could I have gone? I gave birth two weeks early.

They showed me: it was a little girl. 'Natasha,' I said. 'Your dad named you Natasha.' She was healthy enough to look at. Tiny hands and feet. But she had cirrhosis, her liver had had twenty-eight roentgens. And congenital heart disease. Four hours later, they told me my little girl had died. And for a second time, they wouldn't let me have her! What do you mean, you won't give me her! It's me who won't give her to you! You want to take her for science, but I loathe your science! Loathe it! First, your science took him away from me, now it's back for more . . . I won't give her to you! I'll bury her myself. Next to him. (*Her voice drops to a whisper.*)

What I'm telling you, it's not coming out right . . . The words are all wrong. Since the stroke, I'm not meant to shout. And I'm not meant to cry. But I want . . . I want people to know. I've never opened up about all this. When I refused to give them my little girl, our little girl . . . Then they brought me a wooden box: 'She's in there.' I looked inside, and they'd swaddled her. She was lying there all swaddled up. And then I started crying. 'Lay her at his feet. Tell him it's our little Natasha.'

The grave doesn't say 'Natasha Ignatenko'. It only has his name. She still hadn't been named, she had nothing. Only a soul. I buried her soul there.

I always take two bouquets: one for him, and the second one I put on the corner for her. I crawl on my knees at their grave. Always on my knees. (*Incoherently.*) I killed her. I . . . she . . . saved . . . My little girl saved me, she took the whole brunt of the radiation herself, like she was a buffer. So small. Such a teeny little thing. (*Gasping.*) She protected me. But I loved the two of them. Can you . . . Can you kill with love? With such love! Why are they so close? Love and death. They're always together. Who can explain that to me? Who can help me understand? I crawl on my knees at their grave. (*She falls quiet for a long time.*)

In Kiev, they gave me an apartment. In a big block where everyone who left the nuclear plant lives now. We all know each other. It's a big one-bedroom flat, the kind me and Vasya dreamed of. But I went out of my mind there! In every corner, wherever I look: he's there. His eyes. I started redecorating, just so I wouldn't be sitting about, to keep myself occupied. And it's been like that for two years. I've been having this dream. He and I are walking, but he's barefoot. 'Why do you never have shoes on?' 'Because I've got nothing at all.' I went to the church. The priest told me, 'You need to buy some large slippers and put them in somebody's coffin. Write a note that they're for him.' That's what I did. I arrived in

Moscow and went straight to a church. In Moscow, I'm closer to him. That's where he is, in Mitino Cemetery. I told the church man that blah-blah-blah, I need to pass on these slippers. He asked, 'Do you know the way to do it?' He explained it again to me. Just then they brought in an old man for a funeral service. I went up to the coffin, lifted the cloth and put the slippers in. 'And did you write the note?' 'Yes, but I didn't mention which cemetery he was in.' 'They're all in the one world there. They'll find him.'

I had no desire at all to live. At night, I used to stand at the window, staring at the sky. 'Vasya, what should I do? I don't want to live without you.' In the daytime, I'd be passing the kindergarten and would stop and stand. I would have happily looked at the children for hours. I was going crazy! And at night I began asking, 'Vasya, I'd like a baby. I'm frightened of being alone. I can't take it any more, Vasya!' Or another time I asked, 'Vasya, I don't need a man. No one could ever be better than you. But I want a baby.'

I was twenty-five.

I found a man. I told him everything. The whole truth: that I had just one love in my life. I was completely honest with him. We used to meet, but I never invited him home, I couldn't bring him home. Vasya was there.

I worked in pastry. I used to be trimming the sides of a cake, and the tears were rolling down my cheeks. I wasn't crying, but the tears were flowing. All I asked of the girls was: 'Don't feel sorry for me. If you start pitying me, I'll have to leave.' There was no need to feel sorry for me. I had known happiness once.

They brought me Vasya's medal. It was red. I couldn't look at it for long – the tears would flow.

I had a boy. Andrey. Little Andrey. My girlfriends tried to stop me: 'You mustn't have a baby.' And the doctors tried to frighten me: 'Your body won't be able to cope.' And then . . . Then they said he'd be born with one arm missing. His right arm. That's what the screen had shown. 'So?' I thought. 'I'll teach him to write with his left hand.' But he was born normal. A beautiful boy. He's already at school, gets top marks. Now I've got someone who I live and breathe for. The light of my life. He understands me perfectly.

'Mummy, if I go to Granny's for two days, will you be able to breathe all right?' No, I won't! I'm frightened of being apart from him for just a day. We were walking down the street, and I felt myself falling. That was my first stroke. It happened in the street. 'Mummy, should I get you some water?' 'No, you stand right by me. Don't go anywhere.' And I grabbed his arm. I don't remember the rest. Opened my eyes in the hospital. I grabbed Andrey's arm so hard that the doctors could barely unclench my fingers. His arm was blue for ages after. Now, whenever we leave the house, he says, 'Mummy, don't grab me by the arm. I won't leave your side.' He is poorly too: does two weeks at school, then two at home seeing the doctor. That's our life. We worry about each other. And in every corner, there's Vasya. His photos. At night, I talk and talk to him. Sometimes I dream that he's saying, 'Show me our baby.' I bring Andrey, and he leads our little daughter by the hand. He's always with our daughter. He plays only with her.

So that's my life. I'm living in a real and unreal world at the same time. I'm not sure which I like more. (*She gets up and walks to the window.*) There are lots of us here. The whole street. They call it Chernobyl Street. These guys worked their whole lives at the power plant. Many still go there to work shifts, they run the plant with a rotation system now. Nobody lives there any more, and they never will. They've all got serious illnesses, disabilities, but they won't give up their work, they wouldn't even think of it. They'd have no life without the reactor. The reactor is their life. Where else are they needed now? Who needs them? They keep dying. It's a quick death, they die on the go. They'll be walking along and just collapse, black out and never wake up. Bringing flowers to the nurse and their heart fails. Or standing at a bus stop. They're dying, but no one ever really questioned them properly. About what we went through, what we saw. People don't want to hear about death, all these terrible things.

But I've told you about love. About how much I loved.

Lyudmila Ignatenko, wife of Vasily Ignatenko, deceased fireman

The author interviews herself on missing history and why Chernobyl calls our view of the world into question

I am a witness to Chernobyl. The most important event of the twentieth century, despite the terrible wars and revolutions for which that century will be remembered. More than twenty years have passed since the accident, yet I have been asking myself ever since: what was I bearing witness to, the past or the future? It would be so easy to slide into cliché. The banality of horror. But I see Chernobyl as the beginning of a new history: it offers not only knowledge but also prescience, because it challenges our old ideas about ourselves and the world. When we talk about the past or the future, we read our ideas about time into those words; but Chernobyl is, above all, a catastrophe of time. The radionuclides strewn across our earth will live for 50,000, 100,000, 200,000 years. And longer. From the perspective of human life, they are eternal. What are we capable of comprehending? Is it in our power to extract and decipher the meaning of this still unfamiliar horror?

What is this book about? Why have I written it?

This is not a book on Chernobyl, but on the world of Chernobyl. Thousands of pages have already been written on the event itself, hundreds of thousands of metres of film devoted to it. What I'm concerned with is what I would call the 'missing history', the invisible imprint of our stay on earth and in time. I paint and collect mundane feelings, thoughts and words. I am trying to capture the life of the soul. A day in the life of ordinary people. Here, though, everything was extraordinary: both the event itself and the people, as they settled into the new space. Chernobyl for them is no metaphor, no symbol: it is home. How many times has art rehearsed the apocalypse, offered different technological versions of doomsday?

Now, though, we can be assured that life is infinitely more fantastical. A year after the disaster, someone asked me, 'Everybody is writing. But you live here and write nothing. Why?' The truth was that I had no idea how to write about it, what method to use, what approach to take. If, earlier, when I wrote my books, I would pore over the suffering of others, now my life and I have become part of the event. Fused together, leaving me unable to get any distance. The name of my small country, lost in some corner of Europe, which until then the world had heard almost nothing about, now blared out in every language. Our land became a diabolical Chernobyl laboratory, and we Belarusians became the people of Chernobyl. Wherever I go, people eye me with curiosity: 'That's where you're from? What's happening there?' I could, of course, have quickly penned a book (the likes of which have appeared in their dozens) on that night's events at the power plant; on who bears the blame; on how they hid the accident from their own people and from the whole world; on the number of tonnes of sand and concrete needed to build the sarcophagus placed over a reactor spewing death. But something stopped me, something was tying my hands. What was it? A feeling of mystery. At the time, this sudden haunting feeling hung over everything: our conversations, our actions and fears, and it followed in the wake of the event. The monstrous event. A feeling arose in all of us – whether voiced or unvoiced – that we had touched on the unknown. Chernobyl is a mystery that we have yet to unravel. An undeciphered sign. A mystery, perhaps, for the twenty-first century; a challenge for it. What has become clear is that, besides the challenges of Communism, nationalism and nascent religion which we are living with and dealing with, other challenges lie ahead: challenges more fiendish and all-embracing, although still hidden from view. Yet, after Chernobyl, something had cracked open.

The night of 26 April 1986. In the space of one night we shifted to another place in history. We took a leap into a new reality, and that reality proved beyond not only our knowledge but also our imagination. Time was out of joint. The past suddenly became impotent, it had nothing for us to draw on; in the all-encompassing – or so we'd believed – archive of humanity, we couldn't find a key to open this door. Over and over in those days, I would hear, 'I can't find the words to express what I saw and lived through', 'Nobody's ever described anything of the kind to me', 'Never seen anything like it in any book or movie'. Between the time when the disaster struck and when people began talking about it, there was a pause. A moment of muteness we all remember. Somewhere high up, decisions were made, secret instructions were written, helicopters were launched into the skies, vast numbers of vehicles were put on the roads, while down below we waited in fear for reports, we lived on rumours, and yet nobody spoke about the one big issue: what had really happened? Unable to find the words for these new feelings and emotions, unable to find emotions for these new words, we no longer knew how to express ourselves; but we were gradually immersed in the atmosphere of a new way of thinking, and so it has become possible today to pinpoint our state at the time. The truth is that facts alone were not enough; we felt an urge to look behind the facts, to delve into the meaning of what was happening. The effect of the shock. I was searching for those shocked people. They were speaking in new idioms. Voices sometimes broke through from a parallel world, as though talking in their sleep or raving. Everybody near Chernobyl began to philosophize. They became philosophers. The churches filled up again with people – with believers and former atheists. They were searching for answers that could not be found in physics or mathematics. The three-dimensional world came apart, and I have not since met anyone brave enough to swear again on the bible of materialism. We were dazzled by infinity. The philosophers and writers fell silent, derailed from the familiar tracks of culture and tradition. What was most interesting of all in those early days was not talking with the scientists, not with the officials or the high-ranking military men, but with the old peasants. They lived without Tolstoy and Dostoevsky, without the Internet, yet their minds somehow made space for the new picture of the world. Their consciousness did not crumble. Perhaps we would have coped better with a military nuclear crisis like Hiroshima. In fact, that was what we had been drilled for. But the accident happened at a civilian nuclear facility, and we were men and women of our times who believed, as we had been taught, that Soviet nuclear power stations were the most reliable in the world: so reliable, you could even build one in Red Square. Military nuclear power meant Hiroshima and Nagasaki, whereas peaceful nuclear power meant an electric light in every home. Nobody had guessed yet that military and peaceful nuclear power were in fact twins. Accomplices. We grew wiser, the whole world grew wiser, but only after Chernobyl. Today, like living black boxes, Belarusians are recording information for the future. For everybody.

I've spent a long time writing this book. Almost twenty years. I met and talked with former workers at the power plant, with scientists, doctors, soldiers, displaced people and people who returned to their homes in the Zone. With people whose world was Chernobyl; for them, everything inside and out was poisoned, not just the land and the water. They told their stories, searched for answers. Together we pondered. Often they were in a hurry, afraid that time would run out; I didn't recognize that the price of their witness would be their lives. 'Record this,' they would say. 'We didn't understand everything we saw, but let's leave this behind. Someone will read it and make sense of it. Later, after we're gone.' And they had good reason to hurry: many of them are no longer alive. But they managed to send a message.

Everything we know of horror and dread is connected primarily with war. Stalin's Gulags and Auschwitz were recent gains for evil. History has always been the story of wars and military commanders, and war was, we could say, the yardstick of horror. This is why people muddle the concepts of war and disaster. In Chernobyl, we appear to see all the hallmarks of war: hordes of soldiers, evacuation, abandoned houses. The course of life disrupted. Reports on Chernobyl in the newspapers are thick with the language of war: 'nuclear', 'explosion', 'heroes'. And this makes it harder to appreciate that we now find ourselves on a new page of history. The history of disasters has begun. But people do not want to reflect on that, because they have never thought about it before, preferring to take refuge in the familiar. And in the past. Even the monuments to the Chernobyl heroes look like war memorials.

My first journey into the Zone.

The orchards were blossoming, young grass sparkling joyfully in the sun. Birds were singing. Such a profoundly familiar world. My first thought was: everything here is as it should be and carrying on as usual. Here was the same earth, the same water and trees. And their shapes, colours and scents were eternal. It was in nobody's power to alter a thing. But on the first day, I was warned: don't pick the flowers, don't sit on the ground, don't drink the water from the spring. Towards evening, I watched the cowherds trying to drive their weary cattle into the river, but the cows approached the water and turned straight back. Somehow they could sense the danger. And I was told the cats had stopped eating the dead mice, leaving them strewn over the fields and yards. Death lurked everywhere, but this was a different sort of death. Donning new masks, wearing a strange guise. Man had been caught off guard, he was not ready. Ill-prepared as a species, our entire natural apparatus, attuned to seeing, hearing and touching, had malfunctioned. Our eyes, ears and fingers were no longer any help, they could serve no purpose, because radiation is invisible, with no smell or sound. It is incorporeal. All our lives, we had been at war or preparing for war, we were so knowledgeable about it – and then suddenly this! The image of the adversary had changed. We'd acquired a new enemy. Or rather enemies. Now we could be killed by cut grass, a caught fish or game bird. By an apple. The world around us, once pliant and friendly, now instilled fear. Elderly evacuees, who had not yet understood they were leaving forever, looked up at the sky: 'The sun is shining. There's no smoke or gas, nobody is shooting. It doesn't look like war, but we have to flee like refugees.' A world strange yet familiar.

How can we make sense of where we are? What is happening to us, right here and now? There is no one to ask.

In the Zone and around it, there were astonishing numbers of military vehicles. Soldiers were marching with brand-new assault rifles. In full combat gear. For some reason, what stuck in my mind was not the helicopters and armoured personnel carriers, but those assault rifles. The weapons. Men with guns in the Zone. Who were they meant to fire at, defend themselves against? Physics? Unseen particles? Gun down the contaminated earth or the trees? The KGB were working at the power plant. They were looking for spies and saboteurs. The accident was rumoured to be a Western intelligence operation designed to undermine the Socialist order. We needed to stay vigilant.

It was a picture of war. This culture of war crumbled before my eyes. We entered an opaque world where evil offered no explanation, would not reveal itself and did not know the rules.

I watched people transform from their pre-Chernobyl selves into Chernobyl people.

More than once – and this is something to think about – I have heard people say that the behaviour of the firemen extinguishing the fire at the power station on the first night, and the behaviour of the clean-up workers later, resembled suicide. Collective suicide. The clean-up workers often did the job without protective clothing, unquestioningly heading into places where even the robots were malfunctioning. The truth about the high doses they were receiving was concealed from them, yet they were compliant, and later even delighted with the government certificates and medals awarded to them just before they died. Many did not survive that long. So what are they: heroes or suicides? Victims of Soviet ideology and upbringing? For some reason, as the years go by, it is being forgotten that they saved their country. They saved Europe. Just imagine for a moment the scene if the other three reactors had exploded . . .

They were heroes. Heroes of the new history. Sometimes compared to the heroes at the Battle of Stalingrad or the Battle of Waterloo, but they were saving something greater than their homeland. They were saving life itself. Life's continuity. With Chernobyl, man imperilled everything, the whole divine creation, where thousands of other creatures, animals and plants live alongside man. When I visited the clean-up workers, I heard their stories of how – first on the scene and for the first time ever – they dealt with the new human yet inhuman task of burying earth in the earth, meaning they buried in concrete bunkers contaminated layers of soil, along with their entire populace of beetles, spiders and maggots. Insects whose names they didn't even know or couldn't remember. They had an entirely different understanding of death, encompassing everything: from the birds to the butterflies. They were already living in a completely different world – with a new right to life, new responsibilities and a new sense of guilt. Their stories continually featured the idea of time. They were constantly saying, 'the first time', 'never again', 'forever'. They remembered driving through the deserted villages, occasionally meeting some solitary old men who hadn't wanted to leave with the others or had later returned from some unfamiliar places. Those men would sit in the evenings around a rushlight. They mowed with a scythe, reaped with a sickle, chopped down trees with an axe, turned in prayer to the animals and spirits. To God. Just like 200 years earlier. While somewhere high above spacecraft were flying. Time had bitten its own tail, the beginning and end had merged. Chernobyl, for those who were there, did not end in Chernobyl. They were returning not from war, but almost from another world. I realized that they were consciously converting their suffering into new knowledge, donating it to us. Telling us: mind you do something with this knowledge, put it to some use.

The monument to Chernobyl's heroes is the man-made sarcophagus in which they laid to rest the nuclear fire. A twentieth-century pyramid.

In the land of Chernobyl, man's plight makes you sad, but the plight of the animals is even more pitiful. I'll explain. After the humans had gone, what was left in the dead zone? The old graveyards and the so-called bio-burial sites: the cemeteries for animals. Man saved only himself: everything else he betrayed. Once the villages were evacuated, units of armed soldiers and hunters came in and shot the animals. The dogs ran over to the sound of humans. So did the cats. The horses could not understand what was happening. They were in no way to blame – neither the beasts nor the birds, yet they died silently, which was even worse. There was a time when the Mexican Indians, and indeed our own Slav ancestors in pre-Christian Rus, would ask for forgiveness from the animals and birds they killed for food. In ancient Egypt, animals had the right of complaint against humans. In a papyrus preserved in a pyramid, it is written that no complaint by any bull has been found against N. Before departing for the Kingdom of the Dead, the Egyptians would recite a prayer containing the statement: 'I hurt no creature, deprived no animal of grain or grass.'

What has the Chernobyl experience taught us? Has it turned us towards this silent and mysterious world of those other beings?

Once I saw the soldiers go into an abandoned village and begin shooting.

The helpless cries of the animals. They were shrieking in all their different languages. This was written about in the New Testament. Jesus Christ comes to the Temple in Jerusalem and sees animals prepared for ritual sacrifice: with throats slit, dripping blood. Jesus cries out: 'My house shall be called the house of prayer; but ye have made it a den of thieves.' He might have added, 'Ye have made it a place of carnage.' To my mind, the hundreds of bio-burial sites left in the Zone are like pagan temples. Only, to which gods were these sacrifices being offered? The God of Science and Knowledge, or the God of Fire? In this sense, Chernobyl has surpassed the camps of Auschwitz and Kolyma. It has gone beyond the Holocaust. It proposes finitude. It leads to a dead-end.

With fresh eyes, I look at the world around me. A little ant is crawling on the ground, and now it is closer to me. A bird flies in the sky, and it too is closer. The distance between us shrinks. The previous chasm is gone. All is life.

Something an old beekeeper said remains in my memory. I have heard the same thing from others. 'In the morning, I went out into the garden and something was missing, the usual sound was gone. Couldn't hear a single bee – not one! Eh? What was that about? And they wouldn't fly out the second day. Nor the third. Later, they told us there was an accident at the power plant, which wasn't far off. But for a good while we didn't know. The bees knew, but we didn't. From now on, if anything happens, I'll keep an eye on them. On how they live.' Here is another example. I started chatting with some anglers on the river. They told me, 'We were waiting for them to explain it on the TV. For them to tell us how to keep safe. But the worms, just ordinary worms, they buried themselves deep in the ground, a good half a metre or one metre down. We couldn't make sense of it. We kept digging and digging, but couldn't find a single worm for our fishing.'

Who was here first? Who is the stronger and more enduring on the earth: us or them? We could learn a thing or two from the animal kingdom about survival. About how to live too.

Two disasters coincided: a social one, as the Soviet Union collapsed before our eyes, the giant Socialist continent sinking into the sea; and a cosmic one – Chernobyl. Two global eruptions. The first felt closer, more intelligible. People were wrapped up in their day-to-day world of what to buy and where to go. What to believe. Which banner to rally round next. Or whether we should now learn to live for ourselves, find our own lives. The idea was unfamiliar to us, we had never lived like that and did not know how to set about it. Each and every one of us was going through this dilemma. But we would have liked to forget Chernobyl, because our minds just wanted to capitulate. It was a cataclysm for our minds. The world of our beliefs and values had been blown apart. Had we conquered Chernobyl or understood it fully, we would have spent more time thinking and writing about it. Thus we've ended up living in one world, while our minds remain stuck in another. Reality slips away; our consciousness doesn't have room for it.

That's right. We can't catch up with reality.

Here is an example. We're still using the old concepts of 'near and far', 'them and us'. But what do 'near' and 'far' actually mean after Chernobyl, when, by day four, the fallout clouds were drifting above Africa and China? The earth suddenly became so small, no longer the land of Columbus's age. That world was infinite. Now we have a different sense of space. We are living in a space that is bankrupt. What is more, over the last hundred years people have begun to live longer, yet our lifespan is still tiny compared to the life of the radionuclides that have settled on our land. Many of them will live for thousands of years. We can't dream of even a glimpse of such a distant future! In their presence, you experience a new sense of time. And this is all Chernobyl, its imprint. The same thing is happening to our relationships with the past, science fiction, knowledge. The past has proved impotent, and all that is left of knowledge is an awareness of how little we know. We are going through an emotional retuning. Instead of the usual words of comfort, a doctor tells a woman whose husband is dying, 'No going near him! No kissing! No cuddling! This is no longer the man you love, it's a contaminated object.' Here, even Shakespeare bows out, even Dante. The question is whether to go near or not. To kiss or not. One of the heroines of my book went near and kissed, and remained by her husband's side until his death. She paid for it with her health and the life of their baby. But how can you choose between love and death? Between the past and an unfamiliar present? Who could presume to judge the wives and mothers who did not sit with their dying husbands and sons? Next to those radioactive objects. In their world, love has changed. And death too.

Everything has changed, except us.

It takes at least fifty years for an event to become history, but here we have to follow the trail while it is still fresh.

The Zone. It is a world of its own. First it was invented by science-fiction authors, then literature gave way to reality. We cannot go on believing, like characters in a Chekhov play, that in a hundred years' time mankind will be thriving. Life will be beautiful! We have lost that future. A hundred years on, we have had Stalin's Gulags and Auschwitz. Chernobyl. And September 11 in New York. It is hard to comprehend how all this could happen within one generation, within the lifetime of my father, for example, who is now eighty-three years old. Yet he survived it!

What lingers most in my memory of Chernobyl is life afterwards: the possessions without owners, the landscapes without people. The roads going nowhere, the cables leading nowhere. You find yourself wondering just what this is: the past or the future.

It sometimes felt to me as if I was recording the future.

I

Land of the Dead

Monologue on why people remember

I've got a question too. Something I can't really answer myself.

You've decided to write about it. And I don't really want people to know all those things about me, what I went through there. See, I feel the urge to open up, to unburden myself, but then I feel like I'm baring my soul, and that's something I don't want to do.

Remember *War and Peace*? After the war, Pierre Bezukhov is so shaken that he feels he and the whole world can never be the same. But, soon enough, he catches himself slipping back into his old ways: having a go at the coachman, grumbling and growling. So why do people remember things? Is it to get at the truth? For the sake of justice? To let go and forget? Because they realize they were part of some monumental event? Or are they taking refuge in the past? And then there's the fact that memory is fragile, fleeting, it isn't precise facts, it's your conjecture about your own self. It's just emotions, not proper knowledge.

I got all worked up, rummaged through my memory and it came back.

For me, my most terrible time was in childhood. It was the war.

I remember us boys playing 'Mummies and Daddies': we used to undress the tiny children and lay them on top of each other. They were the first children born after the war. The whole village knew what their first words were, which of them had started walking, because children got forgotten in the war. We were waiting for new life to appear. 'Mummies and Daddies' we called that game. We wanted to see new life appearing. We were only eight or ten years old ourselves.

In the bushes by the river, I saw a woman killing herself. She took a brick and was bashing herself on the head. She was pregnant by a Nazi collaborator the whole village hated. When I was still a boy, I saw kittens being born, helped my mother pull the calf out of a cow, took our sow for mating with a boar. I remember . . . remember when they brought my father home dead. He had a sweater on that Mother had knitted him. He must have been shot by a machine gun or assault rifle and some bloody lumps were bulging out from the sweater. He lay on our only bed, we had nowhere else to put him. Then we buried him in front of the house. The earth in his grave wasn't light and soft, it was heavy clay from the beetroot patch. There was fighting all around us. Dead horses and people lying in the streets.

For me, those memories are so off limits that I've never spoken aloud about them.

Back then, I looked at death the same way I looked at birth. It brought up pretty much the same feelings as when the calf came out of the cow. Or the kittens were being born. Or when the woman was killing herself in the bushes. Somehow it all seemed the same thing, no different. Birth and death.

I remember from my childhood the smell in the house when a pig was slaughtered. You've barely nudged me, but I'm sliding back into that nightmare, that horror. Falling headlong . . .

Something else I remember is the women taking us little ones to the bathhouse. Many of them, my mother too, had their wombs slipping out of place (we knew all about it), and they trussed themselves up with rags. I saw it. Their wombs were slipping out from all the hard labour. There were no men: they were all being wiped out at the front or in partisan fighting. There were no horses either, so the women drew the ploughs themselves. They ploughed the vegetable plots and the collective-farm fields. When I grew up and had intercourse with women, I remembered what I saw in the bathhouse.

I wanted to forget. Forget everything. And I was forgetting. I thought the worst was behind me, the war years, and that now I was safe. Protected by my knowledge, by what I'd gone through there. But . . .

I went into the Chernobyl Zone. Been there many times now. It was there I realized I was helpless. And I'm falling apart because of this helplessness. Because I can no longer recognize the world. Everything has changed. Even evil is different. The past can't protect me any more. It can't comfort me, can't offer me any answers. It used to have answers. (*He becomes pensive.*)

Why do people remember the past? Well, I've spoken to you now, put it into words, made sense of something. I don't feel quite so isolated now. But how is it for other people?

Pyotr S., psychologist

Monologue on how we can talk with both the living and the dead

A wolf came into the yard in the night. I looked out the window and it was standing there, eyes blazing. Like headlamps.

I've grown used to it all. Seven years I've been living alone. It's seven years since everybody left. At night, sometimes I'll sit till dawn, just thinking and thinking. Spent the whole of last night hunched up on the bed, then went out to see the sun. What can I tell you? The only righteous thing on the face of the earth is death. No one has ever bribed their way out of that. The earth takes us all: the good, the evil and the sinners. And that's all the justice you'll find in this world. I've slogged my guts out honestly all my life, lived with a clear conscience, but not much justice has come my way. God must have been doling out everyone's share, and when my turn came the pot was empty. For the young, there's a chance death might come knocking, but for us old ones it's a sure thing. We're none of us immortal, not even the tsars or merchants. At first, I was waiting for everyone, thinking they'd all be back. They weren't leaving forever, it was just for a while. But now I'm waiting for death. Dying might not be difficult, but it's scary. There's no church, and the priest doesn't come to these parts. There's nowhere to take my sins.

The first time they told us we'd got radiation, we thought it was some sort of disease, anyone who caught it would drop dead. No, they said, it's something that lies on the ground and gets right inside the ground, but can't be seen. The animals can probably see it and hear it, but people can't. But that's not true! I saw it. This caesium was lying in my vegetable plot until it got wet in the rain. Sort of inky blue, it was. It lay there shimmering in these little lumps. I'd just run back from the collective-farm field and gone to my vegetable plot. And there it was, this blue lump, and a couple of hundred metres away, there was another, as big as the scarf on my head. I yelled to my neighbour and the other women and we all ran around looking for them. Checked all the vegetable plots and the nearby field, a good two hectares. We found maybe four large pieces. One was red. The next day, it was pouring with rain, right from early in the morning. By lunchtime, they were gone. The police came, and there was nothing to show them. We could only describe it all. This big, they were. (*She measures with her hands.*) Like my headscarf. Dark blue and red ones.

We weren't all that scared of the radiation. If we hadn't seen it or known about it, maybe we'd have been frightened, but once we'd had a look, it wasn't really so scary. The police and soldiers stencilled some numbers. By one of the houses, somewhere in the street, they wrote, '70 curies', '60 curies'. We'd been living on our potatoes, our spuds, forever, and here they were saying we couldn't eat them! And they wouldn't let us have onions or carrots, either. You don't know whether to laugh or cry. They told us to work on the vegetable plots in gauze masks and rubber gloves. And to bury the ashes from the stove in the ground. Well I never! And then we had some big scientist come to speak at the village club, says we should wash all the firewood. Come off it! Ordered us to launder all the bed linen and curtains. But they're inside the house! In the wardrobes and linen chests. How could radiation get inside the house? What, past the windows? Past the doors? Barmy! Go and find it in the forests and fields. They locked all the wells shut, covered them in plastic sheeting. The water's 'dirty', they tell us. What do they mean, 'dirty'? It's as clean as clean can be! They came out with some right rubbish. 'You're all going to die.' 'You have to leave.' 'We've got to evacuate you.'

People were scared. Had one heck of a fright. Some folks began burying their valuables in the night. I stashed away my clothes, my certificates of merit for my hard work, and the savings I was keeping for a rainy day. It was so sad! My heart was bursting with sadness! Swear on my life I'm telling you the truth! Then I heard how in one village the soldiers evacuated the people, but an old couple managed to stay behind. The day before they rounded people up and brought the buses over, that couple got their cow and took to the forest. Sat it out, there. Just like in the war, when the Germans burned the villages down. Where's it come from, all this misery? (*Crying.*) It's fragile, our life. I'd rather not cry, but I can't help it.

Hey! Look out the window: a magpie's come. I don't shoo them away. Though sometimes I get the magpies thieving eggs from the barn. All the same, I don't shoo them away. We're all suffering from the same trouble these days. I don't shoo anybody away! Yesterday, I had a hare run in here.

Now if only I had folks in the house every day. There's a woman lives not far from here, in the other village, all alone like me. Told her to move over here, I did. She could help me with things, maybe not with everything, but at least there'd be someone to talk to. To invite in. During the night, I'm aching all over. Get these twisty pains in my legs, like a tingly feeling, it's the nerve shifting about. So I grab something. A fistful of grain. Crunch, crunch . . . and the nerve calms down. I've worked myself to the bone over the years, had my fill of sorrows. Seen enough, I have, don't need anything more. If I died, it would come as a rest. There's no telling how my soul would take it, but my body would be at peace. I've got daughters, I have, and sons. They're all in the town. But I'm not budging from this place! God has granted me long years, though He didn't give me a good lot in life. I know us old ones are a right bother, the children put up with it for a while, then they snap and blurt out something hurtful. Your children only bring you joy while they're little. Our women who moved to the town, they're all in tears. One day they're upset by their daughter-in-law, the next day by their daughter. They want to move back. My good husband, he's lying in the graveyard. If he wasn't lying there, he'd be somewhere else. And I'd be there with him. (*Suddenly, she livens up.*) But why leave? Everything's fine here! It's all lush and blooming. From the gnats to the beasts – everyone's alive and well.

I'll remember it all for you. There were aeroplanes flying non-stop. Every day. They were flying really low over our heads. Flying to the plant, they were. The nuclear reactor. One after the other. And we had the evacuation, they were doing the resettlement. Storming the houses. People were locking themselves in, hiding away. There were cattle mooing and children crying. Like in the war! And the sun kept shining. I plonked myself down and wouldn't go out of the house, though I didn't lock the door. The soldiers knocked: 'Ready to go, missus?' I ask, 'You going to tie up my hands and feet, take me by force?' They were silent for a bit and then left. So young, just children, really! The old women were crawling in front of their houses on their knees. Praying. The soldiers grabbed them by the armpits, lifted them up, one by one, and hauled them into the bus. But I warned them, touch me or use force and I'll clobber you with my cane. Oh, I had a right go at them! Swore at them! I didn't cry. Didn't shed one teardrop that day.

So I sat indoors. There was screaming. Screaming! And then it went quiet. All died down. That first day, I didn't leave the house.

They say that people were marched off in file. And the cattle were marched off too. Like in the war!

My good husband liked to say that man pulls the trigger, but God carries the bullet. We all get our different fates. Some of the youngsters who left have already died in their new place. But here I am, walking with my cane. Still on my feet. When it all gets too dreary, I'll have a cry. The village is empty, but there are all kinds of birds flying here. And the elk wander about calm as anything. (*She cries.*)

I'll remember it all for you. People moved out, but the cats and dogs stayed behind. The first few days, I went and poured them milk, gave each dog a piece of bread. They were standing out in their yards, waiting for their owners. They waited so long. The hungry cats were eating cucumbers and tomatoes. Before the autumn, I mowed the neighbour's grass in front of her gate. Her fence fell down, I nailed it back up. I was waiting for everyone. My neighbour had a little dog; Beetle, he was called. 'Beetle,' I said, 'if you see people first, give me a shout.'

At night, I dreamed I was being evacuated. This officer was shouting, 'Hey, missus, any moment now we're going to burn your place down and bury it. Come on out!' And they took me away to some strange place. I couldn't make sense of it. It wasn't the town or the village. And it wasn't on earth.

Here's something that happened. I had a good little cat; Vaska was his name. In the winter, these hungry rats invaded, there was no hiding from them. They got under the blankets, gnawed a hole through a barrel of grain. And Vaska came to the rescue. Without Vaska, I'd have died. We used to chat, have our lunch together. And then Vaska disappeared. Maybe he was attacked and eaten by the hungry dogs? They were all running around famished till they dropped dead; the cats were so hungry they were eating their own kittens, not in the summer, but they did in the winter. Lord, have mercy! And one woman was gnawed to death by rats. In her own house. Ginger rats, they were. I'm not saying it's true or not, but that's what they say. The tramps come and snoop around here. In the early years, there was good stuff for the taking: shirts, cardigans, fur coats. Could help yourself and off to the flea market. The tramps liked a good drink and sing-song. They'd swear their heads off. One came off a bicycle and fell asleep in the street. In the morning, they found a couple of bones and the bike. True or not? Who knows, but that's what they say.

You get everything living here. The lot! There are lizards, the frogs croaking. Worms wriggling. And there are mice. Everything! It's good in the spring. I love it when the lilac is in flower. The smell of the bird-cherry blossom. While I was firm on my legs, I used to go out for bread, fifteen kilometres each way. When I was a young thing, I'd have flown it. I was used to long walks. After the war, we used to trudge to the Ukraine for seeds. A good thirty or fifty kilometres. People would take sixteen-kilo sacks, while I carried three times that. But these days, I sometimes have trouble crossing the room. Even lying on the stove in summer won't warm an old woman. The police drive over and check up on the village, and they bring me bread. Only what's there to check up on? Just me and the cat here. A new cat. The police toot their horn, me and the cat are so happy. We'll run over. They'll bring him some bones. And they'll ask me, 'What if you're attacked by bandits?' 'Hardly rich pickings for them here, eh? What will they take? My soul? That's all I've got, my soul.' They're good lads. They like to laugh. Brought me some batteries for the radio. I listen to the radio now. I love Lyudmila Zykina, but these days she doesn't sing much. She must have grown old, like me. My good husband liked to say, 'The ball is over, put the violins away!'

I'll tell you how I found the cat. My Vaska was gone. I waited a day, a second day, a month. So I really was left all alone. No one to speak to at all. I walked about the village, through other people's gardens, calling: 'Vaska, Murka, Vaska! Murka!' At first, there were lots of them running about, but then they disappeared. They died out. Death isn't fussy, the earth will take anyone. So I walked and walked. Two days I was calling and calling. On the third day, I find this one, sitting just outside the shop. We look at each other. He's happy, I'm happy. Only he didn't say a word. 'Come on, then,' I tell him, 'let's go home.' He just sits there. 'Meow.' I start coaxing him, 'You want to stay here all on your own? The wolves will get you. They'll rip you to pieces. Come on, I've got eggs and pork fat.' Now how are you to get it across? The cat doesn't understand people's language, so how could he know what I was saying? I walk ahead, and he's running behind me. 'Meow.' 'I'll cut you some fat.' 'Meow.' 'You and I, we'll live together.' 'Meow.' 'I'll call you Vaska.' 'Meow.' And we've already seen through two winters together.

At night, I had this dream someone was calling me. It was the neighbour's voice. 'Zina!' There's silence. And again, 'Zina!'

When it all gets too dreary, I'll have a cry.

I go to the graveyard. My mum is there. My little daughter. She died of typhus in the war. We brought her to the graveyard, buried her, and just then the sun came out from behind the clouds. It was shining so brightly I felt like going back and unburying her. My good husband is there, Fyodya. I'll sit with them all and sigh. You can talk with both the living and the dead. Makes no difference to me, I hear them all. When you're alone. And when you're sad. Terribly sad.

The teacher, Ivan Gavrilenko, lived right by the graveyard. He left to join his son in Crimea. Behind him was Pyotr Miussky's place, the tractor driver. A Stakhanovite model worker. They all used to dream of being Stakhanovite workers. Was a wizard with his hands, could whittle down wood into lace. The finest house in the village, a real beauty! Oh, it was such a shame, made my blood rise when they brought it down. They buried it. The officer shouted, 'Don't grieve, mother. The house is on a hotspot.' He was drunk. I went up, and he was crying. 'Go away, mother! Go!' Chased me away. And past there, it was Misha Mikhalyov's place, he stoked the boilers on the farm. Misha didn't last long. Straight after he left, he died. Behind him was the house of the livestock specialist, Stepan Bykhov. Burned down! At night, some wicked people set fire to it. Outsiders, they were. And Stepan didn't live long either. He's buried near Mogilyov, where his children live. It's a second war. The number of people we've lost! Vasily Kovalyov, Anna Kotsura, Maxim Nikiforenko. We had good fun in the old days. Singing and dancing in the holidays, accordion music. And now it's like a prison. Sometimes I'll close my eyes and walk through the village. How can we have radiation here, I tell them, when there's this butterfly flying, that bumblebee buzzing? And my Vaska catching the mice. (*She cries.*)

So, my love, have you understood my sadness? Pass it on to the people, though I might not be around by then. They'll find me in the earth. Under the roots.

Zinaida Yevdokimovna Kovalenka, returnee

Monologue on a whole life written on a door

I want to testify.

It happened ten years ago, and every day it's still happening to me now. Right now. It's always with me.

We lived in Pripyat. That same town the whole world knows about now. I'm not a writer, but I am a witness. Here's how it was, from the very beginning.

You're living your life. An ordinary fellow. A little man. Just like everyone else around you – going to work, coming home from work. On an average salary. Once a year, you go on holiday. You've got a wife, children. A normal sort of guy. And then, just like that, you've turned into a Chernobyl person. A curiosity! Some person that everyone shows interest in, but nobody knows much about. You want to be the same as anyone else, but it's no longer possible. You can't do it, there's no going back to the old world. People look at you through different eyes. They ask you questions. Was it terrifying? Tell us about when the reactor was on fire. What did you see? Can you still, you know, have children? So your wife hasn't left you? In the beginning, we all turned into some kind of rare exhibits. Just the word 'Chernobyl' still acts like an alarm. They all turn their heads to look at you. 'Oh, from that place!'

That's what it felt like in the first days. We lost not just a town but a whole life.

On the third day, we left our home. The power plant was on fire. Something one of my friends said stuck in my mind: 'You can smell the reactor.' The smell was indescribable. But everyone has read about that in the papers. They turned Chernobyl into a factory of horror stories or, rather, cartoons. But it needs to be understood, because we have to live with it. I'll tell you just my own story.

This is what happened. They announced on the radio: you can't bring any cats with you! My daughter was in tears; she was so afraid of losing her beloved cat that she began stammering. Right, let's put the cat in a suitcase! But the cat wouldn't go in, she was struggling to get out. She gave us all a good scratch. We weren't allowed to take any belongings. Right, I won't, but there's just one thing I will take. Just one! I needed to remove the door to our apartment and take it with us, I couldn't leave the door behind. I would board up the entrance.

That door was our talisman. An heirloom! My father lay on that door. I'm not sure where the custom comes from – they don't do it everywhere – but in our parts, according to my mother, the dead have to be laid on the door from their home. They lie on it until the coffin is brought. I sat the whole night with my father, and he lay on that door. The house was open all night. And that same door is covered in notches, right to the top. It was marked as I grew: a notch for first grade, second grade, seventh grade. One from just before I left for the army. And next to that, you can see my son growing up. And my daughter. Our whole life is written on that door, like on an ancient papyrus. How could I leave it behind?

I asked our neighbour for help; he had a car. He tapped his finger on the side of his head, indicating, 'You have a screw loose, my friend', but I took it. The door. One night. On a motorbike, along the forest road. I took it two years later, when our apartment had already been looted. Picked clean. The police were chasing after me: 'Stop or we'll fire! Stop or we'll fire!' They took me for a looter, of course. It's like I was stealing my own front door.

I sent my wife and daughter to the hospital. They had black spots spreading all over their bodies. They'd spring up and then fade away. The size of an old five-kopeck piece. But nothing was hurting. They were checked. I asked, 'What's the result then?' 'That's not your concern.' 'So whose concern is it, then?'

At the time, everyone around was saying we were all going to die. By the year 2000, the Belarusians will have died out. My daughter had turned six. On the very day of the accident. When I put her to bed, she'd whisper in my ear, 'Daddy, I want to live, I'm only little.' I didn't think she'd understand anything. Whenever she saw a nurse in a white coat at the kindergarten or a cook in the canteen, she'd go crazy. 'I don't want to go to hospital, I don't want to die!' She couldn't stand anything white. We even changed the white curtains in our new place.

Can you imagine seven bald girls together? There were seven of them in the ward. No, that's it! I can't go on! Talking about it gives me this feeling ... Like my heart is telling me: this is an act of betrayal. Because I have to describe her as if she was just anyone. Describe her agony. My wife came back from the hospital. Her nerves snapped: 'If only she'd die, rather than going through this torture. If only I could die, so I wouldn't have to see this.' No, that's it! I can't go on! It's too much. No! ...

We put her on the door. On the door my father once lay on. Until they brought the little coffin. It was so tiny, like the box for a large doll. Like a box.

I want to testify: my daughter died from Chernobyl. But they want us to keep quiet. 'It hasn't been scientifically proved,' they say. 'There isn't enough data. We'll need to wait hundreds of years.' But my human life, it's too short. I can't wait that long. Write it down. You record it at least. My daughter's name was Katya. My little Katya. She was seven years old when she died.'

Nikolai Fomich Kalugin, father

Monologue of a village on how they call the souls from heaven to weep and eat with them

Village of Bely Bereg, Narovlya District, Gomel Province. Speakers: Anna Pavlovna Artyushenko, Yeva Adamovna Artyushenko, Vasily Nikolaevich Artyushenko, Sofia Nikolaevna Moroz, Nadezhda Borisovna Nikolaenko, Alexander Fyodorovich Nikolaenko and Mikhail Martynovich Lis.

Ah, we've got guests. Good people. Didn't have a hunch about a meeting, didn't see no signs. Sometimes my palms will itch, and then somebody will turn up. But today, didn't get a hunch at all. Only sign was the nightingale singing all night, meaning a sunny day ahead. Oh, all the women will be here in a trice. There's Nadya, already hurrying over.

We survived everything, pulled through it all.

*

Oh, I don't want to remember it. Dreadful. They turned us out, the soldiers did. We were swamped by army vehicles and self-propelled guns. One old man had already taken to his bed. He was dying. Where was he meant to go? 'I'll just get up,' he says, crying, 'and walk over to the graveyard. On my own two legs.' What did they pay us for the houses? *How* much? See how gorgeous it is here! Who's going to pay us for all that beauty? It's a holiday spot here!

The place was buzzing with aeroplanes and helicopters. There were KamAZ trucks with trailers, soldiers. Aha, I thought, we must be at war. With the Chinese or the Americans.

My good husband got home from the collective-farm meeting and says, 'Tomorrow we're being evacuated.' And I say, 'But what about the potatoes? We haven't dug them up.' The neighbour knocked on the door and joined my husband for a drink. They had a few and started cursing the chairman: 'We won't go, and that's that. We survived the war, and here we've just got some silly radiation.' We'd rather crawl into the ground. We're not going!

At first, we thought we'd all die in two or three months. They were frightening us, urging us to go. Thank God, we're still alive!

Thank God! Thank God!

No one knows what the next world will be like. It's better in this one. Everything is familiar here. As my mum always said, 'Smarten yourself up, enjoy yourself and do as you please.'

We would go to church and say our prayers.

We were leaving. I took some earth from Mum's grave in a little pouch. Knelt there for a bit. 'Forgive us for leaving you.' At night, I went to her and didn't feel afraid. People were writing their surnames on the houses. On the logs, the fences, the tarmac.

*

The soldiers killed the dogs. They shot them. Bang, bang! Ever since, I can't bear the sound of animals howling.

I used to be a foreman. Worked here forty-five years. I felt sorry for the people. We took our flax to the show in Moscow, the collective farm sent us. I came back with a badge and a certificate of merit. They treated me with respect here, it was all 'Vasily Nikolaevich, our dear Nikolaevich'. And who would I be in the new place? Just some old grandad. This is where I'll lay down and die, the women will fetch me water, warm up the house. I felt sorry for the people. In the evenings, the women used to sing on their way back from the fields, and I knew they weren't getting paid a penny. Just some ticks in their workbooks. And they were singing away.

In our village, the people live together. As one community.

I had this dream when I was already living in the town with my son. This dream that I was waiting for death, waiting for the end. I was instructing my sons: 'When you carry me to our graveyard, I want you to stand with my coffin by the family house, if only for five minutes.' And I was watching from above as my sons carried me there.

It may be poisoned with radiation, but this is my home. There's nowhere else we're needed. Even a bird loves its nest.

I'll finish the story. I was living with my son up on the sixth floor. I'd walk over to the window, look down and cross myself. Thought I could hear a horse. Or a cockerel. And I felt so sad. Sometimes I'll dream of our yard: tethering the cow, and milking her, milking her. Then I wake up, and don't want to get out of bed; I'm still back there. Some of the time I'm here, some of the time there.

By day we lived in the new place, but at night we went back home. In our dreams.

*

In the winter, when the nights are long, sometimes we sit counting everyone who's died. In the town, there are lots who died of nerves and grief at just forty or fifty – is that the right age to go? But we're still living. Every day, we pray to God, ask Him for just one thing: health.

As they say, the place you were born is where you belong.

My good husband was laid up for two months. He wasn't speaking, wouldn't answer me. Like he was upset. I'd potter about the yard, pop back indoors. 'How are you feeling, husband?' He'd look up at the sound of my voice, and I'd feel better. So he might be lying there in silence, but at least he was with me in the house. When someone is dying, you mustn't cry. You'll disrupt their death, make it harder for them. I got a candle from the cupboard and put it in his hands. He took it and was breathing. I saw his eyes misting over. I didn't cry. Just asked for one thing: 'Say hello over there to our little daughter and my precious mum.' I prayed to be together with him. Some people's prayers are answered, but He hasn't granted me death. I'm still here.

I'm not afraid of dying. Nobody gets to live twice. Look how the leaves blow away, the trees topple down.

Don't cry, old girls! We were star workers for all those years, Stakhanovite workers. We survived Stalin, survived the war! If we hadn't laughed and had fun, we'd have hanged ourselves ages ago. Two Chernobyl women are chatting. One says, 'Have you heard, everyone's got the white blood cancer now?' The other says, 'Rubbish! Yesterday, I cut my finger and the blood was red.'

Home is where the heart is. The sunshine isn't the same anywhere else.

My mother once told me: take an icon, turn it back to front, then leave it like that for three days. No matter where you are, you'll find your way home. I had two cows and two heifers, five pigs, some geese and chickens. And a dog. I clasped my head in my hands and paced about the orchard. There were so many apples! Everything's lost. Damn, it's all gone!

I washed the house, whitewashed the stove. You have to leave bread on the table and salt, a bowl and three spoons. As many spoons as there are souls in the house. To be sure you'll return.

All the hens' combs were black, not red: that was the radiation. And we couldn't make cheese. We went a month without soft cheese or hard. The milk wouldn't sour, it curdled into lumps, these white lumps. It was the radiation.

Had that radiation stuff in my vegetable plot. The whole plot went white, completely white, like it was dusted with something. With some little specks. I thought maybe something had been carried from the forests. The wind had sprinkled it.

We didn't want to leave. No, we didn't! The men were drunk, throwing themselves in front of the cars. The officials were going from house to house, trying to persuade people. The orders were, 'Leave all your belongings behind!'

The cattle went three whole days without water and food. Took them to be slaughtered! A newspaper journalist arrived: 'How are you feeling? How are things going?' The drunken milkmaids nearly murdered the fellow.

The chairman and some soldiers were hovering round my house. They tried frightening me: 'Come out or we'll set fire to the place! Hey, pass us the petrol can!' I started running about – grabbed a towel, a pillow . . .

Now you tell me how that radiation works, according to science. Tell us the truth, because we'll be dying soon in any case.

*

And you reckon they don't have it in Minsk, seeing as it's invisible?

My grandson brought me a dog. Called it Radium, because we're living in this radiation. So where did my Radium get to? Always at my feet. I'm frightened he'll run out of the village and the wolves will get him. I'll be left all alone.

In the war years, all through the night, the guns would thud and chatter away. We built dugouts in the forest. They kept on bombing. They burned down everything, not just the cottages, but even the vegetable plot and the cherry trees.

So long as there's no war . . . I'm terrified of war!

On Radio Yerevan, a caller asks: 'Is it okay to eat Chernobyl apples?' The answer: 'Yes, but bury the cores deep in the ground.' A second caller asks: 'What is seven times seven?' The answer: 'Ask a Chernobyl survivor, they'll count it on their fingers.' Ha ha.

They gave us a new little house. Made of stone. You know what, in seven years, we haven't hammered in a single nail. It's a foreign land! Everything's foreign. My good husband cried and cried. He would work all week, driving the tractor on the collective farm, waiting for Sunday; and when Sunday came, he'd lie there facing the wall and crying.

Nobody can trick us again, we're not budging from this place. We've got no shop, no hospital. There's no light. We sit around paraffin lamps and rushlights. But we're happy! We're home.

In the town, my daughter-in-law followed me round the apartment with a cloth, wiping the doorknobs, the chairs. And everything was bought with my money, all the furniture, the Lada. When the cash runs out, nobody needs you.

*

Our children took the money, and inflation ate up the rest. The money they gave us for the smallholding, for the cottages, the apple trees.

On Radio Yerevan, a caller asks: 'What is a Radio Nanny?'* 'A grandmother from Chernobyl.' Ha ha.

I was walking for two weeks. With my cow. People wouldn't let me into their houses. I had to sleep in the forest.

They're frightened of us. We're infectious, they say. What is God punishing us for? He's angry? We're not living like humans, not living by God's laws. We're killing each other. That's why He's angry.

My grandchildren came in the summer. The first few years they didn't come; like everyone else, they were scared. But now they visit, and they take produce home, they'll pack whatever you give them. 'Grandma,' they said. 'Have you read a book about Robinson Crusoe?' He lived alone, just like us. Without anyone. I brought half a sack of matches with me, an axe and a spade. And now I have pork fat, eggs, milk – all my own produce. Just one thing you can't plant, and that's sugar. Here you've got all the land you could want! Plough a hundred hectares if you like. And there's no authorities. Nobody bothering you here. No higher-ups. We're free.

We returned along with our cats. And dogs. We came back together. The soldiers and riot police wouldn't let us in, so we came by night. Took the forest footpaths. The partisan paths.

There's nothing we need from the state. We grow everything ourselves. All we ask is to be left alone! We don't need any shops or buses. We go twenty kilometres on foot for our bread and salt. We can fend for ourselves.

*

* A popular Soviet radio programme for children.

A whole band of us returned. Three families. But the place was gutted: they'd smashed the stove, taken the doors and windows. The floors. The bulbs, switches, sockets – they'd unscrewed the lot. Not a living thing left. I fixed it all up again with my own hands. Oh yes!

The wild geese are screeching: spring is here. It's time to sow. And here we are in our empty houses. Only the roofs are sound.

The police used to shout at us. They'd drive over, and we'd hide in the forest. Like hiding from the Germans. Once they turned up with the prosecutor. He was threatening to take us to court. I said, 'They can throw me in jail for a year, but the moment I'm out, I'll be straight back.' Their job was to yell, ours was to keep quiet. I have a medal for being one of the top combine drivers, and here he was threatening me. Said I'd get sent down for Article 10. As if I'm a criminal.

Every day, I dreamed about my house. I was returning home: one time I'd be digging in the vegetable plot, another time making the bed. And I'd always find something: a shoe, or some chicks. All good omens, signs of happiness to come. Of a homecoming.

At night, we plead with God, by day we plead with the police. You ask me why I'm crying. I don't know why. I'm happy to be living in my own dear home.

We survived everything, pulled through it all.

I'll tell you a joke. The government issues an edict about benefits for Chernobyl victims. Anyone living within twenty kilometres will be addressed as 'O Beaming One'. Anyone within ten kilometres will be addressed as 'O Radiant One'. And anyone right near the plant who survived, 'O Luminous One'. See, O Radiant One, we're alive. Ha ha.

I finally reached the doctor. 'My legs won't carry me, love. And my joints are aching.' 'Get rid of the cow, old girl. The milk is poisoned.' 'Oh, I can't do that,' I said, crying. 'What with my legs aching and my knees hurting, I'm not giving the cow away. She keeps me fed.'

I've got seven children, all of them living in the town. I'm here on my own. Whenever I start missing them, I'll sit next to their photos, have a chat. I do everything alone. Painted the house on my own, laid on six tins of paint. That's how I live. Raised four sons and three daughters. And my husband died early. I'm on my own.

I ran into a wolf: he was standing bang in front of me. We stared at each other, and he leapt off to the side. Darted away. My hair went stiff with fear.

All wild animals are afraid of man. Leave the animals alone and they'll steer clear of you. Before, if you were walking in the forest and heard voices, you'd run over to them, but now people will hide from each other. Heaven forbid you ever meet a man in the forest!

What's written in the Bible is all coming true. In the Bible it says about our collective farm. And about Gorbachev. It says there'll be a big leader with a mark on his forehead, and a great power will crumble to dust. And then the Day of Judgement will come. Those in the towns will all die, and just one man will be left in the villages. People will be happy to find a human footprint! Not a human being, just a footprint.

We use lamps for light. Paraffin ones. Ah, the women have already told you. When we kill a pig, we'll carry it down to the cellar or bury it in the ground. The meat will last three days in the ground. We make moonshine from our own grain. From jam.

I've got two sacks of salt. We'll be all right without the state! Got plenty of firewood – surrounded by forest. The house is warm. The lamps are lit. All good! I keep a nanny goat, a billy goat, three pigs and fourteen hens. There's land and grass to your heart's content.

Water in the well. Freedom! We like it! What we have here is no collective farm, it's a commune. Communism! We'll buy another horse. And then we won't need anyone. Just one horse.

We didn't just move home, as one shocked journalist put it, we moved a hundred years back in time. Reaping by sickle, mowing by scythe. We thresh the grain with a flail right here on the tarmac. My good husband makes baskets. And in the winter, I do embroidery and weaving.

In the war, our family lost seventeen members. Two of my brothers were killed. Mum cried and cried. An old woman was going begging from village to village. 'You're in mourning?' she asked Mother. 'Don't grieve. The one who lays down his life for others is a holy man.' I'd do anything for the Motherland. The only thing I couldn't do is kill. I'm a teacher, and I taught that we should love one another. Good will always triumph. Children are just little, their hearts are pure.

Chernobyl. The war to end all wars. There's nowhere to hide. Not on land, in water or in the skies.

First, the radio was turned off. We don't know any of the news, but it's a quiet life. We don't get upset. People come here and tell us there are wars everywhere. And they say Socialism is finished, we're living under capitalism. And the tsar will return. Is it really true?

Sometimes a boar will come into the orchard from the forest, sometimes an elk. People come rarely, though. Just the police.

Come and visit my home.

And mine. It's so long since I had guests in the house.

Dear Lord, I cross myself and pray! Twice the police smashed up my stove with an axe. They took me away on a tractor. But I came back! If they'd let people in, everyone would come crawling home on their knees. The news of our troubles has spread across the world. Only the dead are allowed back. They are brought here. But the living come in the night. Through the forest.

They're all longing to come here for Radunitsa. To the last man. Everyone wants to pray for their dead. The police will let in those who are on their lists, but no children under eighteen. People get here, and they're so happy to stand near their house, near an apple tree in the orchard. First they cry at the graves, then they go to their old houses. And there they cry some more and pray. Light some candles. They lean against their fences as if they were graves. They might put a wreath by the house, hang a white towel over the gate. The priest will read a prayer: 'Brothers and sisters! Have patience!'

They take white loaves and eggs to the cemetery, many bring pancakes instead of bread. Whatever they've got to hand. Everyone sits at their loved ones' graves. They call out, 'Sister, we're here to visit you. Come and eat with us.' Or, 'Dearest Mum, dearest Dad.' They call the souls down from heaven. People whose loved ones have died during the year will cry, while those who lost them earlier won't. They have a chat, bring up memories. Everyone prays. Even those who don't know how to pray join in.

You mustn't weep for the dead at night. Once the sun's gone down – you mustn't. May the Lord rest their souls. Grant them the kingdom of heaven!

Laugh and the world laughs with you. There's a Ukrainian woman sells big red apples at the market. She was touting her wares: 'Come and get them! Apples from Chernobyl!' Someone told her, 'Don't advertise the fact they're from Chernobyl, love. No one will buy them.' 'Don't you believe it! They're selling well! People buy them for their mother-in-law or their boss!'

*

One of the locals got back from jail. It was under a prisoner amnesty. He lived in the next village. His mum died, and they buried the house. He washed up on our shores. 'Give us a hunk of bread and some pork fat, missus. I'll chop your firewood for you.' He goes begging.

The country's a mess – and people are coming here to escape. Some folks are fleeing from people, others from the law. And they live here on their own. They're not from these parts. They're grim, no light in their eyes. They get drunk and set fire to the houses. At night, we sleep with pitchforks and axes under the bed. In the kitchen, there's a hammer by the door.

In the spring, a fox with rabies was running about. When they have rabies, they're gentle as can be. They can't stand the sight of water. Put a pail of water out in the yard – and fear not! It will run away.

They've started coming here. Making movies about us, though we never get to see the films. We've got no TV or electricity. All we've got is the window to look through. And prayer, of course. We used to have Communists instead of God, but now there's just God left.

We were honoured citizens. I was a partisan, spent a year with the resistance. And when our side pushed back the Germans, I found myself on the front line. Wrote my surname on the Reichstag: Artyushenko. Hung up my army coat and built Communism. And where is it now, our Communism?

We've got Communism here. We live as brothers and sisters.

When the war broke out, there were no mushrooms or berries that year. Would you believe it? The earth itself could feel trouble brewing. 1941. How I remember it! Oh, I haven't forgotten the war. A rumour went round that our prisoners had been brought here, and if you found a relative, you could take him home. Our women upped and ran off to meet them! That evening, some brought home their loved ones while others brought unknown men. But there was a bastard in our midst. Lived just like the rest of us, married with two kids. He went to the commandant's office and informed them that some of the men we'd taken were Ukrainians. We had Vasko, Sashko . . . The next day, the Germans came on their motorbikes. We fell on our knees and begged them. But they led them out of the village and gunned them down with their rifles. Nine men. So young, such lovely guys! Vasko, Sashko . . .

So long as there's no war . . . I'm terrified of war!

The officials would come, they'd shout their heads off, but we acted deaf and dumb. And we survived everything, pulled through it all.

And I'm lost in my own thoughts. Thinking and thinking. At the graveyard. Some of them are wailing loudly, some softly. Others might be chanting, 'Open up, yellow sands. Open up, dark night.' Well, you can call people back from the forest, but not from the yellow sand. I'll talk to him lovingly: 'Ivan, Ivan. What am I to do with my life?' But he doesn't say a thing, whether kind or harsh.

As for me, I'm not frightened of anyone: not the dead, not the wild animals, nobody. My son comes from the town and pesters me. 'You're all on your own here. What if someone comes and strangles you?' And what would he take from me? There's just my pillows. In my little hovel, pillows are all the finery you'll get. The moment the bandit climbs in, he'll stick his head through the window, and I'll lop it off with an axe. Maybe there is no God, maybe it's somebody else, but up above us, there's someone there. And I am alive.

In the winter, an old man hung up a calf's carcase he'd cut up in the yard. Just then, they came along with some foreigners. 'What are you doing, old man?' 'Letting the radiation out.'

It really happened, people told us about it. One man buried his wife and he was left all alone with their baby boy. Took to drink from grief. He used to take the wet things off the tot and put them under the pillow. And the wife – it could have been her, could just have been her soul – would come at night and she'd wash, dry and fold the things up. Once he caught sight of her. Called out – but she just vanished. Into thin air. Then the neighbours told him: the moment you spot so much as a shadow, lock the door with the key, then maybe she won't run away so fast. But she didn't come back again. Now what was that all about? Who was it that was coming?

Don't you believe me? Then tell me, where do these tales come from? Maybe it really happened? Oh, you educated types . . .

Why did that Chernobyl blow up? Some say it's the scientists to blame. Trying to catch God by the beard, and He had the last laugh. And it's us that suffer!

We've never had it easy. It's never been calm. Right on the eve of the war, they were taking people in. Capturing them. Took three of our men. Came in their black cars and took them out of the fields and we never saw them again. We've always lived in fear.

I don't like crying. I like hearing new jokes. They grew some tobacco in the Chernobyl Zone. In the factory, they made it into cigarettes. Each pack had a message: 'Ministry of Health: smoking is bad for you. This is your *last* warning.' Ha ha. But our old fellows are all smokers.

The one thing I've got left is my cow. I'd happily give her away if it would mean there was no war. I'm terrified of war!

The cuckoos are calling, the magpies chattering. Roe deer are running about. But nobody can say if they'll carry on multiplying. One morning, I looked into the orchard and there were boars grubbing about. Wild boars. You can resettle people, but not the elk and the boars. And the water takes no notice of boundaries, it flows where it will, over the ground, under the ground.

*

A house can't exist without people. Wild animals need people too. Everybody is looking for people. A stork came. A beetle climbed out. It all brings me joy.

Everything's hurting, old girls. Oh, how it hurts! You have to be gentle. You carry a coffin gently. Carefully. No banging it against the door or the bed, no touching it against anything or knocking it. Or you'll bring bad luck – you can expect another death. May the Lord rest their souls. Grant them the kingdom of heaven! And the spot where you're buried, that's where they'll wail. Here we've got nothing but graveyards. Graves all around. Tipper trucks droning, and bulldozers. The houses are falling down, the gravediggers never stop work. They've buried the school, the village soviet, the bathhouse. Our whole world, and the people aren't the same. There's one thing I don't know: does a person have a soul? What's it like? And how do all of them fit into the world to come?

For two days, my grandad was dying. I hid behind the stove, watching to see how it would fly out of him. I went to milk the cow. Ran back indoors and called him. He was lying there, with his eyes open. Had his soul flown away? Or was there nothing? And if not, how will we ever meet up again?

The priest says we're immortal. We say our prayers. O Lord, give us the strength to bear the trials of our lives.

> *Monologue on how happy a chicken would be to find a*
> *worm. And what is bubbling in the pot is also not forever*

My first fear . . .

The first fear fell from the sky. It floated down with the water. But some people, and there were quite a few, were as cool as stone. I swear on the Cross! When they'd had a few drinks, the older guys liked to say, 'We marched to Berlin and won the war.' They said it like they had you pressed up against the wall. They were the victors! They had the medals to prove it.

My first fear came in the morning when we found suffocated

moles in the orchard and vegetable plot. Who had choked them? They don't usually come above ground. Something must have driven them out. I swear on the Cross!

My son rang from Gomel. 'Do you have any cockchafers flying about?'

'No cockchafers here, not even the grubs. They've gone into hiding.'

'What about earthworms?'

'It only takes an earthworm to make a chicken happy. They've all gone too.'

'No beetles or worms is the first sign of high radiation.'

'What's radiation?'

'Mum, it's a kind of death. Talk Dad into leaving. You can stay with us for a bit.'

'But we haven't planted the vegetable plot . . .'

If everyone was smart, where could we find a fool? Okay, so it was on fire. Fires don't last long. No one was frightened at the time. We didn't know about all that atomic stuff. I swear on the Cross! We were living right beside an atomic power station, thirty kilometres as the crow flies, or forty by road. We were very pleased. You could buy a bus ticket and just go there. The town had shopping as good as Moscow's – there was cheap sausage, always meat in the shops. You could choose. Good times!

But now there's nothing but fear. Folks say the frogs and midges will live on, but not the humans. Life will go on without humans. They tell all these tall tales. Anyone fond of them tales is a fool! But there's no smoke without fire. We've been hearing it for a long time now.

I'll turn on the radio. They keep on and on about radiation. But we're better off with radiation. I swear on the Cross! Just look: they've brought in oranges, three sorts of sausage, there you go! In our little village! My grandchildren have travelled half the planet. The youngest girl came back from France, that's where Napoleon once marched from. 'Granny, I saw a pineapple!' The second grandchild, her little brother, was taken to Berlin for treatment. That's where Hitler came barging in from. In their tanks. It's a new world now. Everything is different. Is it the radiation to blame, or who is it? And what's that stuff like? Maybe they've shown it in the movies? Have you seen it? Is it white, what does it look like? What colour? Some say it's got no colour or smell, but others say it's black. Like the earth! If it's no colour, then it's like God. God is everywhere, but you can't see Him. They're trying to frighten us! But we've got apples hanging in the orchard, and leaves on the trees, potatoes in the field. I don't believe there ever was any Chernobyl, they made it all up. Tricked people. My sister and her man left. Didn't move far, just twenty kilometres away. They'd been there two months, when a neighbour comes running: 'Your cow's radiation has got on to ours. The cow keeps falling down.' 'And how did it get on to her?' 'It flies around in the air, like dust. It can fly.' Stuff and nonsense! But here's something true. My grandad had bees, he had five hives. Well, for three days they wouldn't fly out, not one bee. They were sitting in the hives. Waiting it out. My grandad was running about the yard: what kind of disaster was this? What the Devil was up? Something had gone wrong with nature. And as our neighbour, who's a teacher, explained to us, their system is cleverer than ours, because they could feel it right away. The radio and the papers still weren't saying anything, but the bees knew. They only flew out on the fourth day. The wasps . . . We had wasps, a nest of them over the porch, nobody touched them; then suddenly the next morning they were gone, no sign of them dead or alive. They came back six years later. Radiation. It frightens people and animals alike. And birds too. And even the trees are scared, but they can't talk. They can't tell you. But the Colorado beetles carry on crawling, same as before, eating our spuds, gobbling up every last leaf. They are used to poison. Just like us.

Come to think of it, in every house someone has died. The street across the river, all the women there live without menfolk; there are no men, the men have died. On our street, my grandad lives here, and one other man. God takes the men first. Why's that? Nobody can decipher that for us, nobody knows that secret. But think about it: if only the men were left, and no women, it wouldn't be much better. They drink, my love, they drink. The sadness drives them to it. Who feels happy about dying? When a person dies, they get so sad! There's no comforting them. Nobody can make them feel better, there's nothing you can do. They drink and natter. Talk about things. They get drunk, make merry and poof! – they're gone. Everyone dreams of an easy death. But how can you earn it? The soul is the only living matter. My dear girl. And our women are all barren, their women's parts have been cut out, a good one in three. Young and old alike. Not all of them had managed to have children in time. When I think of it, it's all just flown past.

What can I add? You just need to live. That's all.

And another thing. Before, we used to churn our own butter, sour our own cream, make soft cheese, hard cheese. We cooked milk soup with dumplings. Do they eat that in the towns? You pour water into some flour and mix it up into these ragged pieces of dough, then drop them in a pot of boiling water. You boil them a bit and add some milk. Our mum explained it to us, showed us how. 'Learn how to make this, children. It's what I learned from my mother.' We drank birch drink and maple drink. We steamed runner beans in iron pots in the big oven. Made kissel from the cranberries. And in the war, we picked nettles, orache and other wild greens. Swelled up from hunger, but we didn't die. Had berries and mushrooms in the forest. But now we've got a life where that's all ruined. We thought it would last and last, things would carry on the way they'd always been. And what was bubbling in the pot would be there forever. Never would have believed it could all change. But that's what's happened. You're not allowed milk, not allowed beans. No mushrooms, no berries. They tell you to soak the meat for three hours. And you have to drain off the water twice when you boil potatoes. But you can't fight against God. You just need to live.

They frighten us that our water can't be drunk. But how can you go without water? Everyone has water inside. There's nobody without water in them. You even find water in stones. Well, this is water we're talking about, maybe it's eternal? The whole of life comes from it. Who can you ask? No one will tell you. And you pray to God, you don't ask Him things. You just need to live.

And now the wheat's coming up. It's good wheat.

Anna Petrovna Badaeva, returnee

Monologue on a song without words

I'm on my knees, begging you.

Find Anna Sushko for us. She used to live in our village, in Kozhushki. Her name is Anna Sushko. I'll give you her description, and you print it. She's got a hump, been mute since childhood. Lived alone. She was sixty. In the resettlement, they put her in an ambulance and took her away to an unknown destination. She never learned to read or write, so we've never had a letter from her. They carted off people living on their own and the disabled to homes. Hid them away. Nobody knows the addresses. You please print it all.

The whole village cared about her. We looked after her like she was a little child. Someone would chop her firewood, someone else would bring her milk. Someone would sit with her in the evenings, light the stove. It's been two years since we stopped drifting about those strange places and came back home. And tell her that her house is sound. The roof is still there, and the windows. We'll help her fix what was smashed up and looted. Just give us the address where she's living in her misery, we'll go and fetch her. Bring her back. Before she dies of sadness. I'm begging you, on my knees. There's an innocent soul suffering out in that strangers' world.

There's another detail. I forgot. When something hurts, she'll start warbling this song. No words, just her voice. She can't talk. When she's in pain, she'll sing it: 'Ah-ah-ah.' Whimpering.

'Ah-ah-ah . . .'

Maria Volchok, neighbour

*Three monologues on ancient fear, and on why one man
stayed silent while the women spoke*

> *The K. family. Mother and daughter, and a man
> (the daughter's husband) who did not say a word.*

Daughter:

At first, I cried night and day. All I wanted was to cry and speak.
We are from Tajikistan, from Dushanbe. There's a war there.

I shouldn't really talk about it. I'm expecting, pregnant. But I'll
tell you. In the daytime, they'll come on the buses, checking passports. Ordinary men, but with guns. They look at the passports
and throw some men off the bus. And right by the doors, they'll
shoot them. They don't even bother taking them aside. I would
never have believed it, but I saw it. I saw them taking two men,
one was so young, good-looking, he was shouting something to
them. Both in Tajik and in Russian. He was saying that his wife
had just had a baby, and he had three small children at home. And
they just laughed, they were young as well, really young. Ordinary
men, but with guns. He fell to the ground, he was kissing their
trainers. Everyone was silent, the whole bus. The moment we
drove off: 'Rat-a-tat.' I was afraid to look back. (*She cries.*)

I shouldn't really talk about it. I'm expecting. But I'll tell you.
Just one thing: please don't use my surname, my first name is Svetlana. We still have family there. They'll kill them. I used to think
we'd never have another war again. It was our huge country, we
loved it. The strongest! Before, they used to tell us that our life in
the Soviet Union was poor and modest because we'd been through
a great war, the people had suffered, but now we had a powerful
army, no one could touch us. No one could conquer us! But then we
started shooting each other. It's a different kind of war from before.
Our grandfather told us about the old war. He reached Germany,
got to Berlin. Now it's neighbour shooting neighbour, boys who
went to school together, and now they're killing each other, raping
the girls they sat next to at school. Everyone has gone crazy.

Our husbands won't speak. The men are just silent, they won't say a word to you. People shouted to their backs that they're running away like women. Cowards! Betraying the Motherland. But what have they done wrong? Is it your fault if you can't shoot? Or you don't want to? My husband is Tajik, he was supposed to join the war and kill. But he said, 'Come on, we're leaving. I don't want a war. Don't need a gun.' He loves carpentry, caring for horses. He doesn't want to shoot. That's just what his heart is like. He doesn't like hunting either. It's his land, they speak in his language, but he left. Because he didn't want to kill other Tajiks just like him. Some person who he doesn't know and who's done nothing against him. He wouldn't even watch TV there, covered up his ears. But he's lonely here. Over there, his brothers are fighting, one has been killed. His mother's still over there, and his sisters. We travelled here on the Dushanbe train, there was barely a window with glass in it, it was freezing, no heating, there wasn't any shooting, but along the way they threw stones at the windows, smashed the glass: 'Russians out! End the occupation! No more robbing us!' And he's a Tajik, and he heard all this. And our children did too. Our little girl had just started school, she was in love with a boy. A Tajik. Comes home from school: 'Mummy, what am I, Tajik or Russian?' You can't explain it to her.

I shouldn't really talk about it, but I'll tell you. The Pamiri Tajiks are fighting the Kulobi Tajiks. They are all Tajiks, they've got the one Koran, the one faith, but Kulobis are killing Pamiris, and Pamiris killing Kulobis. First, they got together in the square, shouting and praying. I wanted to understand it, so I went there too. Asked the elderly men: 'Who are you coming out against?' They answered, 'Against Parliament. We've been told this Parliament is a very bad man.' Then the square emptied, and the shooting started. It suddenly became a different country, one we didn't know. The East! Before, it had felt like we were living in our own land. Under Soviet law. There were so many Russian graves left behind, and no one to weep at them. They were taking cows to graze in the Russian cemeteries. Goats too. Russian old men were wandering about the dumps, picking through the rubbish.

I worked in a maternity hospital. I was a nurse, doing night duty.

A woman was giving birth, she was having a hard time and scream-ing. An orderly came rushing in, wearing unsterile gloves and gown. What had happened? Why was she in the delivery room like that? 'Girls, we've got bandits!' They were in black masks and armed. They ran straight over to us. 'Give us drugs! Surgical alco-hol!' 'There's no drugs, no alcohol here!' They got the doctor up against the wall: 'Give it!' And then the woman giving birth shouted in relief. A happy shout. And a tiny baby began crying, it had just been born. I leant over it, don't even remember if it was a boy or girl. It still had no name, nothing. And the bandits asked us: 'Is it Kulobi or Pamiri?' Not 'boy or girl', but 'Kulobi or Pamiri'. We stayed silent. And they yelled, 'Who is it?' We said nothing. Then they grabbed the tiny baby, it had only been in the world maybe five or ten minutes, and hurled it out the window. I'm a nurse, I've seen children die. But this . . . My heart nearly jumped out of my chest. I shouldn't really remember it. (*She begins crying again.*)

After that incident . . . I had eczema come up on my arms. My veins were bulging. And I felt so numbed, I could hardly crawl out of bed. I'd go towards the hospital and turn back again. And I was expecting a baby myself. How could life go on? How could I give birth there? So we came over here. To Belarus. Narovlya is a small town, it's quiet. Now don't ask me any more. Leave me alone. (*She stops speaking.*) Wait . . . I want you to know. I'm not frightened of God. It's man I'm frightened of. When we first got here, we asked, 'Where's this radiation of yours?' 'It's right where you're standing.' But that's the whole land itself! (*She wipes away tears.*) People left. They were scared.

I don't find it as scary here as it was back there. We're left with-out a homeland, we don't belong anywhere. The Germans left for Germany, the Tatars went to Crimea when they got permission, but nobody needed the Russians. What was there to hope for? What could we expect? Russia never saved her own people, because the country's too big, too endless. To be honest, I don't actually feel that Russia is my Motherland; we were brought up with a different idea: that our Motherland was the Soviet Union. So you don't know any more what's right. Nobody cocking guns here – that's already a good thing. We were allocated a house, my husband was given a job. I wrote to my friends, they got here yesterday. Came for good. They arrived in the evening and were too frightened to leave the station, wouldn't let their children out, just sat on their suitcases. They waited till morning. And then they saw that people were walking about the streets, laughing, smoking. They were shown where our street was and brought to our place. They still weren't themselves, because we'd all forgotten what normal, peaceful life was like. That you can walk down the streets in the evenings. That you can laugh. The next morning, they went to the grocery shop, saw all the butter and cream, and right there in the shop – they described it to us – they bought five bottles of cream and drank them on the spot. People looked at them like they were mad. But they hadn't seen cream or butter in two years. Over there, you couldn't buy bread. It was war. You can't explain it to someone who doesn't know what war is, only knows it from the movies.

My soul was dead there. Who would I give birth to, with my soul dead? There aren't many people here. The houses are empty. We live near the forest. I get frightened when there are too many people. Like at the station, in the war . . . (*She begins sobbing frantically and then becomes quiet.*)

Mother:

The war. It's all I can talk about. Why did we come here? To the Chernobyl Zone? Because no one will kick us out. From this land. It's nobody's, God has taken it over. People have deserted it.

In Dushanbe, I worked as the deputy chief of the railway station, and there was another deputy who was Tajik. Our children grew up together, went to school together, we always sat at the same table for holidays: New Year, May Day, Victory Day. We drank wine and ate pilaff together. He used to call me 'little sister, my Russian sister'. And suddenly he walks over – we shared an office – stops in front of my desk and shouts: 'When are you going back home to Russia? This is our land!'

Just then I thought my head would explode. I jumped up: 'Where's your jacket from?'

'Leningrad,' he said, taken aback.

'Take off your Russian jacket, you bastard!' I said, yanking his jacket off. 'Where's the fur hat from? You were bragging about how it was sent from Siberia! Take off your hat, bastard! And your shirt! Your trousers! They were made in a Moscow factory! They're Russian too!'

I would have stripped him to his underpants. A big hefty guy. I came up to his shoulder, but suddenly – where on earth did I get the strength from? – I would have yanked the lot off him. A crowd had already collected round us. He yelled, 'Get away from me, you madwoman!'

'No, hand over all your Russian stuff! Give it back to me!' I nearly went nuts. 'Take off your socks! Your shoes!'

We were working night and day. The carriages were jammed with people fleeing. Lots of Russians on the move. Thousands! Tens of thousands! Hundreds of thousands! It was like a second Russia. I'd sent out the 2 a.m. train to Moscow, and some children from the town of Kurgan-Tyube were left in the waiting hall, they'd missed the Moscow train. I shut them away, hid them. These two guys come up to me. They've got guns.

'Hey, guys, what are you doing here?' My heart was racing.

'It's your fault, you left all the doors open.'

'I was sending the train off. Didn't have time to close them.'

'Who are those children?'

'They're our local kids, from Dushanbe.'

'Ah, but maybe they're from Kurgan? They're Kulobi?'

'No, no. They're local.'

They left. But what if they'd gone into the waiting hall? They would have killed the lot of us. Including me: a bullet to the forehead! Only one thing rules there: a man with a gun. The next morning, I put the children on the train to Astrakhan, ordered them to be shipped like watermelons, not to open the doors. (*At first, she is silent, then cries for a long time.*) Is there anything more terrifying than man? (*She is silent again.*)

When I was already living here, I'd look over my shoulder every couple of minutes; it felt like someone was behind me, at the ready.

Waiting. Back there, not a day went by without me thinking about death. I always left the house in clean clothes: a freshly washed blouse and skirt, clean underwear. Any moment, I could be killed! But now I can walk through the forest alone, I'm not frightened of anyone. There's nobody in the forest, not a soul. I'll walk and think back: did it really all happen to me, or did I dream it? Another time, I'll run into some hunters: they've got a rifle, a dog and a dosimeter for the radiation. Men with guns again, but this time they aren't chasing after humans. If I hear a shot, I'll know they're shooting the crows or chasing a rabbit. (*She is silent.*) So I'm not scared here. I can't feel afraid of the earth, the water. It's man I'm afraid of. Back there, he can buy an assault rifle in the market for a hundred dollars.

I remember one guy. A Tajik. He was chasing another guy. Hunting a human! The way he was running, the way he was breathing, I knew instantly that he wanted to kill him. But the other guy hid. He escaped. And the man comes back, he's passing me and he says, 'Excuse me, where can I get a drink of water here?' Just asked casually, as if nothing had happened. We had a tank of water at the station, I showed him where. And then I looked him in the eye and said, 'Why are you hunting each other down? Why are you killing?' And he even looked a bit sheepish. 'Hey, lady, don't make trouble.' But when you get them together, they're different. If there were two or three of them, they'd have me up against the wall. When it's one, you can talk to him.

From Dushanbe we got to Tashkent, but we needed to go further: to Minsk. There were no tickets – and nothing you could do! They'd set things up very neatly. You couldn't get a seat on the plane unless you slipped someone a bribe: you'd have endless problems – with the weight, the size; you couldn't take this, had to get rid of that. They sent us twice to the scales, and I still didn't twig. Then I slipped them some cash. 'Now you're talking; see, why make things hard for yourself?' It was that simple! Whereas before, they made us unload our two-tonne container. 'You're from a war zone. You carrying weapons? Hashish?' I went to the head of the airport, and in the waiting room met a good woman. She explained what was what: 'You won't get far here. No point demanding justice, you'll end up with your container dumped in a field and all your stuff looted.' What could we do? We didn't sleep all night, unloading our possessions: clothes, mattresses, old furniture, an old fridge and two sacks of books. 'Valuable books, are they?' We had a look: *What Is to Be Done?* by Chernyshevsky, *Virgin Soil Upturned* by Sholokhov. We laughed. 'And how many fridges?' 'Just the one, and it's already been bashed up.' 'Why didn't you bring the right paperwork?' 'Well, how were we to know? It's our first time fleeing a war.' We lost two motherlands at once: our Tajikistan and the Soviet Union.

I go off on walks in the forest, think about things. The others all crowd round the TV: they want to know what's happening back there. But I don't.

Those were the days. A different life. I was an important person then, with a military rank: lieutenant-colonel of the railway forces. Here, though, I was out of work until the town council took me on as a cleaner. I wash floors. One life is over, and I don't have the strength for another one. Some of the locals feel sorry for us, others resent us. 'Those refugees are stealing our potatoes. They dig them up at night.' My mother used to look back on the war years and say people felt more pity for each other. The other day, they found an abandoned horse in the forest. It was dead. And a dead hare in another spot. They weren't even killed, just dead. Everybody was bothered about it. But when they found a dead tramp, that somehow got ignored.

People have grown used to the human dead everywhere.

*

> *Lena M. is from Kirghizia. Sitting with her on the doorstep of the house, as though posing for a photo, were her five children and their cat, Blizzard, whom they had brought with them.*

We left like we were fleeing a war.

We grabbed our belongings, and the cat followed right behind us, all the way to the station, so we took the cat too. The train trip took twelve days; for the last two days, all we had left were some jars of sauerkraut and boiling water from the urn. We took turns guarding the doors, some guys with a crowbar, some with an axe or hammer. Let me tell you about it. One night, we were attacked by bandits. We were almost killed. These days, people will kill you for your TV set or fridge. We left like we were fleeing a war, though where we lived in Kirghizia, they weren't fighting yet. There'd been a massacre in the city of Osh. Kirghiz and Uzbeks. It all died down quickly, seemed to go quiet. But there was something in the wind, something brewing in the streets. I'm telling you. It wasn't just us Russians, like you'd think, even the Kirghiz were afraid. There were queues for bread, and they'd shout: 'Russians, go home! Kirghizia is for the Kirghiz!' And then shove us out of the queue. They'd add something in Kirghiz, like, there wasn't enough bread for them, let alone to feed us. I didn't really understand their language, just learned some words for haggling in the market.

We used to have a Motherland. It's gone now. Who am I? Got a Ukrainian mother, my dad's Russian. I was born and raised in Kirghizia, then married a Tatar. Who are my children? What's their ethnicity? We're all mixed up, our blood is mixed. In our passports, for me and my children it says 'Russian'; but we're not Russian. We're Soviet! But the country I was born in doesn't exist. The place that we called home and the times which were also our home don't exist now. We've become as homeless as bats. I've got five children: my eldest boy is in the eighth grade at school, the youngest girl is in kindergarten. I brought them here. Our country doesn't exist, but we still exist.

I was born and grew up there. Built a factory, worked in a factory. 'Go back to your own land, everything here is ours.' They didn't let us take anything but the children: 'Everything here is ours.' And where's it all mine? People are fleeing, they're leaving. All the Russian people. The Soviet people. Nobody needs them, there's nobody waiting for them.

I used to be happy once. All my children were born from love. I had them in this order: boy, boy, boy, then girl, girl. I'm not going to say anything else. I'll start crying . . . (*But she adds a few more words.*) This is where we're going to live. It's our home now.

Chernobyl is our home. Our Motherland. (*She suddenly smiles.*) The birds here are the same as we had back there. And there's a statue of Lenin . . . (*Standing at the gate, as she says goodbye.*) Early one morning, in the house next door, they were banging away with hammers, taking the boards off the windows. I met the woman. 'Where are you from?' 'Chechnya.' She never says anything. She walks about wearing a black headscarf.

I meet people, they're amazed, can't understand it. 'What are you doing to your children, you're killing them. You're committing suicide.' I'm not killing them, I'm saving them. Look, forty years old and I'm completely grey. At forty! Once they brought a German journalist to the house, and he asked: 'Would you take your children somewhere there was plague or cholera?' But the plague and cholera are different. This threat here, I don't feel it. I don't see it. It's nowhere in my memory.

It's men I'm afraid of. Men with guns.

Monologue on how man is crafty only in evil, but simple and open in his words of love

I was running. Running from the world. First I hung out at the railway stations. I liked the stations because there were loads of people, yet you were alone. Then I read about it in the newspapers – and came here. There's total freedom here. I'd say it's heaven. There is nobody here, just wild animals wandering around. I live among the animals and the birds. Who can say I'm alone?

I've forgotten my own life. Don't ask me about it. I remember what I read in books and the things other people told me, but I've forgotten my own life. I was a young man. Fell into sin. But there's no sin that the Lord won't forgive if your repentance is sincere. That's it. Men are unjust, but the Lord is infinitely patient and merciful.

Why? There's no answer. Man cannot be happy. He shouldn't be happy. The Lord saw that Adam was lonely and He gave him Eve. So he could be happy, not so he could sin. But man is no good at happiness. I don't like the twilight; that in-between time, like it is

73

now. Going from light to night. If I start thinking, I can't really see where I was before. Where was my life? That's it. I don't really care whether I live or not. Human life is like the grass, it flowers, withers and gets tossed in the fire. I've grown fond of thinking. Here you could just as easily be killed by wild animals as die of the cold. Or die from thinking. There isn't a soul here for dozens of kilometres. You can drive out the demons with fasting and prayer. Fasting for the flesh, and prayer for the soul. But I'm never lonely, a believer can never be lonely. Right. I visit the villages. I used to find macaroni and flour. Sunflower oil and tinned food. But now I go scavenging at the cemetery. They leave out food and drink for the dead, but the dead don't need it, and they won't take offence. In the field, there's wild wheat. Mushrooms and berries in the forest. There is total freedom. I do a lot of reading.

If we turn to the holy pages, the Book of Revelation: 'And there fell a great star from heaven, burning as it were a lamp, and it fell upon the third part of the rivers, and upon the fountains of waters; And the name of the star is called Wormwood: and the third part of the waters became wormwood; and many men died of the waters, because they were made bitter.' I'm trying to fathom that prophecy. Everything has been predicted, it's all written in the holy books, but we don't know how to read. We aren't bright enough. The Ukrainian word for 'wormwood' is 'Chernobyl'. There was a sign for us in those words. But man is too restless. Too vain and petty.

I found this, by Father Sergey Bulgakov: 'as God most certainly created the world, the world cannot altogether fail.' And so we must 'valiantly endure history to the end'. Right. And someone else wrote, I don't remember who, but I remember their idea: 'Evil of itself is not an essence, but the privation of good, just as darkness is no more than the lack of light.' It's easy to come by books here. You won't pick up so much as an empty clay pitcher, no spoons or forks, but there are plenty of books. Recently I found a volume of Pushkin. 'The thought of death delights my very soul.' I remembered that line. That's it. 'The thought of death.' I'm here all alone. Sometimes I think of death, I've grown fond of thinking. The silence helps you prepare yourself. Man lives in the midst of death, but he doesn't understand what death is. I'm here all alone. Yesterday I chased a she-wolf and her cubs out of the school building, that's where they were living.

Here's a question: can the world be captured genuinely in words? It's the word that comes between man and his soul. That's it.

And another thing: the birds, the trees, the ants – they've all become close to me. Before, I didn't have those feelings. Didn't have an inkling of them. Something else I read somewhere: 'The universe above us and the universe below us.' I think about them all. Man is horrifying. And peculiar. But here you don't feel like killing anyone. I go fishing, I've got a rod. That's it. But I don't shoot animals, don't set traps. My favourite literary character, Prince Myshkin, said, 'How can you look at a tree and not be happy?' That's it. I love thinking. But people spend too much time complaining, instead of thinking.

Why look too closely at evil? It's troubling, of course. Sin is not really rocket science. It's important to acknowledge the imaginary. It says in the Bible there's one meaning for the initiates, for the rest there are parables. Take a bird, or some other living creature. We can never understand them, because they live for themselves, not for others. That's it. In short, everything around us is flowing.

All creatures are four-legged, looking down at the earth, drawn towards the earth. Man alone stands up on the earth, and raises his hands and head to the sky. To prayer. To God. The old woman in the church prays: 'Each of us shall reap according to his sins.' But neither the scientists nor the engineers nor the soldiers will admit it. They think, 'I've got nothing to repent of. Why should I repent?' That's it.

My prayer is simple. I say it silently. 'Lord, I cry unto me! Give ear!' Man is crafty only in evil, but he's so simple and open in his plain words of love. Even for philosophers, the word is only an approximation of the thought they have experienced. The word genuinely attunes to what's in our soul only in prayer, and in prayerful thoughts. I can feel it physically. 'Lord, I cry unto me! Give ear!'

And man too.

Man frightens me, but I always like meeting one. A good man. That's it. The only people living here, though, are bandits in hiding and people like me. Suffering souls.

My surname? I haven't got a passport. The police took it. They beat me up. 'What are you bumming about here for?' 'I'm not bumming about, I'm repenting my sins.' They beat me even harder. Punched me in the head. So you can write: Nikolai, a servant of God.

And now a free man.

The Soldiers' Choir

Artyom Bakhtiyarov, private; Oleg Leontyevich Vorobey, clean-up worker; Vasily Iosifovich Gusinovich, reconnaissance driver; Gennady Viktorovich Demenev, policeman; Vitaly Borisovich Karbalevich, clean-up worker; Valentin Komkov, driver, private; Eduard Borisovich Korotkov, helicopter pilot; Igor Litvin, clean-up worker; Ivan Alexandrovich Lukashuk, private; Alexander Ivanovich Mikhalevich, radiation monitoring technician; Oleg Leonidovich Pavlov, helicopter pilot, major; Anatoly Borisovich Rybak, commander of a security platoon; Viktor Sanko, private; Grigory Nikolaevich Khvorost, clean-up worker; Alexander Vasilyevich Shinkevich, policeman; Vladimir Petrovich Shved, captain; Alexander Mikhailovich Yasinsky, policeman.

Our regiment was alerted for duty. We travelled a long time. No one was saying anything concrete; they only told us where we were going once we were at Belorussky station in Moscow. One guy, I think he was from Leningrad, was protesting: 'I don't want to die.' They threatened him with a court martial. The commander told him in front of the whole unit: 'You'll go to prison or face a firing squad.' My feelings were different, quite the opposite. I wanted to do something heroic, put my character to the test. Maybe it was just a boyish impulse? There were guys from the entire Soviet Union serving with us: Russians, Ukrainians, Cossacks, Armenians. It was unsettling, and yet oddly fun.

So they took us there, right up to the power plant. Gave us white coats and white caps; gauze face masks. We cleaned the grounds. It was one day shovelling and scraping below, one day up on the roof of the reactor. Did all the work by spade. The guys up on the roof were called 'storks'. The robots couldn't take it, the equipment was going crazy. But we did our work. We sometimes got blood coming out of our ears, our noses. A tickling in the throat, your eyes stinging. There was this constant drone in your ears. You felt thirsty, but lost all appetite. We weren't allowed to do our morning exercises, to keep us from breathing in extra radiation. Though we travelled to work in open-top trucks.

But we did our work well. And we're very proud of that.

We drove inside. There was a sign saying 'Prohibited Zone'. I've never been at war, but it was sort of a familiar feeling. Like I had it somewhere in my memory. Where from? Something to do with death.

Along the way, we saw dogs and cats that had gone wild. Sometimes they behaved oddly, wouldn't recognize people, ran away from us. I couldn't understand what was wrong with them, until we got the order to shoot them. The houses had been sealed off, the farm machinery was lying there abandoned. It was interesting to look at. No one was there, just us and the police patrolling. You'd go into a house and there would be pictures hanging, but no one inside. There were documents lying around: Young Communist League membership cards, people's ID, certificates of merit. In one house, a TV set was taken – just borrowed for a bit – but I didn't notice anyone taking stuff home. For starters, it felt as though people were about to come back any moment. And also, it was like . . . Something to do with death.

The guys went into the unit, right up to the reactor. To take photos, so they could show off back home. There was fear, but at the same time this overwhelming fascination: we wanted to see what it was like. Me, I refused. I had a young wife, didn't want to risk it, but the others downed 200 mils of vodka each and off they went. Yeah . . . (*After a pause.*) They came back alive, meaning everything was okay.

Then we were out patrolling on night duty. There was a bright moon, like a hanging lamp.

The village street, not a soul . . . At first, there were lights still on in the houses, but then they switched off the electricity. We'd be driving along, and a wild boar would shoot across our path out of the doors of a school. Or a fox. Wild animals were living in the houses, schools and village halls. And there were the posters: 'Our goal is happiness for all mankind', 'The world proletariat will be victorious', 'Lenin's ideas live forever'. There were red flags in the collective-farm offices, all these brand-new pennants, piles of certificates embossed with the profiles of Marx, Engels and Lenin. Portraits of our leaders on the walls, plaster busts of them on the desks. There were war memorials everywhere. I didn't see any other monuments. Just these hastily erected houses, grey concrete cowsheds, rusting hay towers. And then more Mound of Glory war memorials, big and small. 'So that's our life?' I asked myself, looking at everything through fresh eyes. 'That's how we live?' Like some warrior tribe had moved on from its makeshift camp. Hurried off somewhere.

Chernobyl blew my mind. I began thinking.

An abandoned house. All shut up. There was a kitten in the window. I thought it was pottery. I went up: it was alive. It had eaten all the flowers in the pots, the geraniums. How had it got in there? Or had it been forgotten?

A note on the door: 'Dear stranger, do not look for valuables. We never had any. Use whatever you need, but please don't loot. We're coming back.' On other houses, I saw signs in all different colours: 'Forgive us, dear family home!' They had said goodbye to their homes as if they were parting from a person. They wrote, 'We're leaving this morning', or 'Leaving this evening'. They put the date, even the hour and the minute. Notes scrawled in children's handwriting on pages torn from exercise books: 'Don't beat the cat, or the rats will eat everything up.' 'Don't kill our Zhulka. She's a good dog.' (*He closes his eyes.*) I've forgotten it all. I remember that I went, but nothing else. I've forgotten it all. Two years after I was demobbed, something happened to my memory. Even the doctors don't understand it. I can't count out money, I get all mixed up. I'm sent from one hospital to the next.

Have I told you this already? You'd walk over, thinking a house was empty, open the door, and there'd be a cat sitting there. Yeah, and those children's notes.

I got called up for service.

And our duty was not letting the local people back into evacuated villages. We stood in lines near the roads, we built dugouts and look-out towers. For some reason, the locals called us 'partisans'. This was peacetime, but there we stood, decked out in our army gear. The peasants couldn't understand why they weren't allowed, say, to fetch a bucket from their yard, a jug, a saw, an axe. Or get the crops in. How could you explain it? Well, actually, you had soldiers standing on one side of the road not letting anyone in, while on the other side cows were grazing, combine harvesters were whirring away and they were threshing the grain. The women all crowded together and wept. 'Lads, come on, let us in! That's our land, our homes.' They brought us eggs and pork fat, moonshine. 'Let us in . . .' They were crying for their poisoned land, their furniture, all their belongings.

Our duty was to keep them out. An old woman came, carrying a basket of eggs; we had to confiscate and bury them. She'd milked a cow and was carrying a pail of milk. A soldier was with her; the milk had to be buried. If they secretly dug up their spuds, you confiscated them. And the beetroots, the onions, pumpkins: they all had to be buried. Those were the instructions. The harvest was glorious, enviable. And such beauty all around. It was a golden autumn. We all had the faces of madmen, both them and us.

Meanwhile, the papers were trumpeting our heroism. What heroic guys we were, Young Communist League volunteers!

Who were we really? What were we doing there? I'd like to know that, read it. Though I was there myself.

I'm a military man, my duty is to do what I'm ordered. I took an oath.

But that's not all, there was also the heroic urge. They'd nurtured it, sown it in our minds at school, at home. And then the political instructors set to work on us. The radio, the television. Different people reacted differently: some liked the idea of being interviewed, getting a mention in the papers, others looked on it as a job, and there was also a third type. I met them, they lived with the feeling they were doing something heroic, taking part in history. We were paid well, but the question of money didn't really come into it. My salary was 400 roubles, but out there I was on 1,000 roubles in old money. In those days, that was a lot. We were criticized later for raking in the money and returning with hopes of jumping the queue for cars and furniture. That was hurtful, of course. Because there was also that heroic spirit.

Just before we went out there, this fear sprang up. For a short time. But once we got there, it disappeared. If only I could have seen it, that fear. Out there, we just had orders, assignments, work. I was interested in viewing the reactor from above, from a helicopter: I wanted to see what had happened, what it looked like. But that was forbidden. On my card, they wrote, '21 roentgens', but I'm not sure that was really the right figure. The idea was simple enough: you flew to Chernobyl (which, by the way, was a little district town, not something grander, like I'd imagined), there you had a monitoring technician, ten or fifteen kilometres from the station, taking the background radiation readings. Those readings were then multiplied by the number of hours we flew in a day. As for me, I went up in the helicopter and flew to the reactor: there and back, both ways. One day it was 80 roentgens, the next it would be 120. At night, I circled over the reactor for two hours. We were shooting in infrared; on the film, the pieces of graphite looked sort of 'overexposed'. By day, though, you couldn't see them.

I spoke with some of the scientists. One told me, 'I could lick your helicopter with my tongue and nothing would happen to me.' Another said, 'Guys, what are you doing, flying without protection? You want to shorten your lives? You need to clad your helicopter! Plate it!' Well, if you want something done, you've got to do it yourself. We lined the seats with lead, cut chest protectors for ourselves from thin lead sheeting. Though it turned out lead protects you from one kind of ray, but not the other. Everyone's faces went red and burned, we couldn't shave. We were flying from morning till night. There was nothing spectacular, just work. Hard work. At night, we sat in front of the TV, the World Cup was on. All the talk revolved around football, of course.

We began thinking more deeply ... Oh, it would have been maybe three or four years later. When the first man fell ill, then the second. Someone died, another man went mad, another killed himself. That's when we began thinking more deeply. And we'll understand at least something, I reckon, in another twenty or thirty years. I was in Afghanistan (for two years) and in Chernobyl (for three months) – the most vivid moments of my life.

I didn't tell my parents we'd been sent to Chernobyl. My brother just happened to buy a copy of *Izvestia* and saw my picture. He showed it to Mother: 'There, see: a hero!' Mother burst into tears.

We were on our way to the power plant.

These columns of evacuees were moving towards us. They were driving trucks and tractors, herding cattle. Night and day. All this in peacetime.

We were driving along, and do you know what I saw? On the roadsides, in the sunlight, this barely visible sparkling. Some sort of tiny crystal particles glinting. We were driving towards Kalinkovichi, via Mozyr. Something was shimmering in different colours. We all talked about it, we were amazed. In the villages where we were working, we immediately noticed holes burned through the leaves, especially the cherry trees. We picked tomatoes and cucumbers, and there were tiny black holes in the leaves. It was autumn. The redcurrant bushes were bright red with berries, the branches were sagging to the ground with apples – and of course we couldn't resist. We ate them. They'd warned us not to, but we decided to hell with it and ate them.

I went out there. Although I didn't have to. I volunteered. In the early days, I didn't meet anyone half-hearted. It was later they got that blank look in their eyes, once they'd grown used to things.

Svetlana Alexievich

Were we after medals? Special benefits? Rubbish! I personally didn't need anything. An apartment, a car, what else? Ah, yes, a dacha. I had all that. It was the male lust for adventure that kicked in. Real men were going on a real mission. And the others? They could hide behind their mothers' skirts. One guy brought a certificate saying his wife was giving birth, another had a small child. Sure, it was risky; sure, it was dangerous – radiation. But somebody had to do it. And what about our fathers in the war?

We got home. I took everything off, all the stuff I'd been wearing there, and threw the lot down the rubbish chute. I gave the cap to my little son as a present. He kept asking for it. He wore it non-stop. Two years later, he was diagnosed with a brain tumour.

You can write the rest yourself. I don't want to say any more.

I'd just got back from Afghanistan. I wanted to live, get married. Wanted to marry right away.

But suddenly I had a call-up notice, with a red band headed 'Reservist Mobilization': I needed to turn up within the hour at the address given. My mother started crying at once. She decided they were sending me back to war.

Where were we going, and why? The information was hazy. A reactor had blown up. And what of it? In Slutsk they got us changed, handed out uniforms, and revealed that we were on our way to Khoyniki, the district centre. We got to Khoyniki, where the people still knew nothing. Like us, they were seeing dosimeters for the very first time. We were taken further, to a village. They were celebrating a wedding: the young couple were kissing, they were playing music and drinking moonshine. Just your usual wedding. And our orders were to dig out the soil to a spade's depth and chop down trees.

At first, we were issued weapons: assault rifles. In case the Americans attacked. In our political sessions, they lectured us on sabotage operations by Western intelligence, on their covert work. In the evenings, we'd leave our weapons in a separate tent in the middle of the camp. A month later, they took them away. There were no saboteurs: just roentgens and curies.

For 9 May, Victory Day, a general arrived. They lined us up and wished us a happy holiday. One of the men plucked up courage and asked, 'Why are you hiding the background radiation levels from us? What doses are we getting?' There's always one wise guy. So, once the general had left, the commanding officer summoned the guy and gave him an earful. 'Stirring up trouble! You panic-monger!' A couple of days later, we were issued gas masks, but nobody used them. Twice they showed us dosimeters, but they wouldn't let us touch them. Every three months, we could go home for a couple of days. There was one instruction: buy vodka. I came back hauling two rucksacks stuffed with bottles. The guys swung me up into the air.

Before going home, we were all called to see a KGB officer, who strongly advised us never to speak to anyone anywhere about what we'd seen. When I got back from Afghanistan, I knew I'd live! After Chernobyl, the opposite was true: it was when you were back home that it would kill you.

I'm home now. And it is all just beginning.

What stands out, what's burned in my memory?

We spent the whole day running around between villages. With radiation monitoring technicians. And not one woman offered us an apple. The men were less frightened, they brought out the moonshine and the pork fat. 'Come and have lunch.' You felt bad about refusing, but the idea of dining on pure caesium didn't exactly fill you with joy. So you'd down a drink, but no nibbles.

Penny bun mushrooms crunched under the wheels of our trucks. Call that normal? These fat, lazy catfish, six or seven times bigger than usual, were swimming in the river. Call that normal?

In one village, we did sit down for a meal. Of roast lamb . . . The host got drunk and admitted, 'It was a young little lamb. Slaughtered it because I couldn't bear the sight of it. Ugh, what a monstrosity! Puts me off eating it.' I downed a glass of moonshine, after those words. Our host laughed. 'We've adapted here like Colorado beetles.'

We brought the dosimeter up to the house: it was off the scale.

*

Ten years have passed. It already feels like it never happened; if I hadn't fallen sick, I'd have forgotten it all.

You have to serve the Motherland! Serving the Motherland is our sacred duty. I received: underwear, foot cloths, boots, epaulettes, cap, trousers, tunic, belt and rucksack. And I was off! They gave me a tipper truck; I was shifting concrete. I sat in the cab and believed the steel and glass were shielding me. Just get on with it! It'll be all right. We were young guys, unmarried. We didn't take respirators with us. No, actually there was one guy. An older driver. He was always in his mask. But we didn't wear ours. The traffic cops didn't use masks. We were inside our cabs, but they were standing in radioactive dust for eight hours at a time. We all got paid well: three times the usual salary plus a travel allowance. We drank. We knew that vodka helped. It was a first-rate method for restoring the immune system after exposure to radiation. And it helped with the stress. They knew what they were doing with the famous hundred mils of vodka per soldier during the war. It was the usual picture: drunken policemen fining drunken drivers.

Don't write about the miracles of Soviet heroism. They happened, those miracles! But first it was incompetence, sloppiness, and only then came the miracles. Throwing yourselves on to pillboxes, flinging your chests against machine guns. But nobody writes about the fact they should never have issued those orders in the first place. They slung us in like they dumped the sand on the reactor. Like we were sandbags. Every day, they hung up a new soldiers' bulletin: 'Their brave and selfless work . . .' 'We shall stand firm and triumph.' They came up with a lovely name for us: 'Soldiers of the Fire.'

For my heroism, I got a certificate of merit and 1,000 roubles.

At first, it was baffling. It all felt like an exercise, a game.

But it was genuine war. Nuclear war. A war that was a mystery to us; where there was no telling what was dangerous and what wasn't, what to fear and what not to fear. No one knew. And there was no one to ask. There was a genuine evacuation. The stations . . . Oh, the scenes at the stations! We helped shove kids through the windows of carriages; got the queues under control. There was queuing for tickets at the ticket office, queuing for iodine at the pharmacies. People cursed and fought in the queues. They kicked in the doors of the booze shops and kiosks; they smashed the windows and prised out the metal grilles. There were thousands of evacuees. They were living in the village halls, schools, kindergartens. They wandered around half-starved. All their money had run out, and the shops had been picked clean.

I'll never forget the women who laundered our underwear. There weren't any washing machines. No one had thought of that, they hadn't brought any. So they were washing by hand. All the women were elderly. Their arms were blistered and crusting over. Our underwear wasn't just 'dirty', it'd had dozens of roentgens. 'Have some food, lads.' 'Get some sleep, lads.' 'You're only young, lads. Take care of yourselves.' They felt sorry for us and were crying.

Are they alive today?

Every year, on 26 April, we get together, those of us who were there. Those who are still left. We look back on those days. You were a soldier in that war, you were indispensable. All the bad stuff gets forgotten, and that is what stays. What lingers is the fact they couldn't cope without you. You were essential. Our system, our military, operates pretty much superbly in an emergency. Out there, you were finally free and needed. Freedom! At moments like that, the Russian people show how great they are. How special! We'll never be like the Dutch or Germans. And we'll never have good roads or groomed lawns. But we'll always have heroes!

Here's my story.

The call went out, and I answered it. Had to be done. I was a Party member. Communists, forward! Was just the way things were. I was in the police force, senior sergeant. They promised me a new star on my epaulettes. It was June of '87. There was a mandatory medical you had to pass, but I was sent without a check-up. Someone somewhere copped out, as they say, brought a certificate saying he had a stomach ulcer, so I took his place. Urgently. Just the way things were. (*He laughs.*) By that time, the jokes had sprung up. Overnight. A guy comes home from work and complains to his wife, 'They say tomorrow it's off to Chernobyl or hand in my Party card.' 'But you aren't in the Party!' 'That's why I'm trying to get hold of a Party card by the morning.'

We went there as soldiers, but at first they organized us into a brick-laying crew. We were building a pharmacy. I came straight down with this weakness, this sort of drowsiness. At night, I was coughing. I went to the doctor. 'Everything's fine. It's the heat.' They brought meat, milk and soured cream from the collective farm to the canteen, and we ate it. The doctor wouldn't touch a thing. They'd cook the food, and he'd log in his journal that everything was within the limits, but he wouldn't try it himself. We noticed that. Just the way things were. We lads were fearless. The strawberries were ripening, the hives were dripping with honey.

Looters began raiding the place. They were hauling everything off. We boarded up the doors and windows. Sealed up the safes in the collective-farm offices, the village libraries. Then we cut off all the utilities, severed the electricity to the buildings in case of fire.

They ransacked the shops, tore the grilles off the windows. There was flour, sugar, sweets all trodden into the floor. Broken jars. They'd move people out of one village, while five or ten kilometres away people were still living there. Belongings from the deserted village would drift over to them. Just the way things were. We used to keep guard. Along would come the former chairman of the collective farm, with locals who'd been resettled some place, but were returning to harvest the wheat and sow the fields. They were carting off hay bales. Inside the bales, we found they'd hidden sewing machines, motorbikes, TV sets. And the radiation was so strong that the TVs wouldn't work. It was barter: they gave you a bottle of vodka, and you gave them permission to carry away a pram. They sold the stuff, exchanged the tractors and seed drills. One bottle, ten bottles – nobody was interested in money. (*He laughs.*) Proper Communism. Everything had its price: a can of petrol would be half a litre of moonshine, an astrakhan coat was two litres, a motorbike was whatever you could agree on. I did my six months. According to the schedule, we were meant to serve there for six months, then they'd send replacements. We were held back a bit, because the soldiers from the Baltics refused to go. Just the way things were. But I know the place was cleaned out, they made off with everything that could be lifted. They took the test tubes from the school chemistry labs. They brought the Zone back here. You can go look for it in the markets, the second-hand shops, the dachas.

All that remained behind barbed wire was the land. And the graves. Our past, our great country.

We got to our destination. Changed uniforms.

The question was: what were we in for? 'An accident, it happened a while back,' our captain reassured us. 'Three months ago. Nothing to worry about now.' The sergeant said: 'Everything's fine, just wash your hands before eating.'

I served as a radiation monitoring technician. The moment it was dark, these guys would drive over in their cars to our mobile unit. We were offered money, cigarettes, vodka to let them rummage through the confiscated items. They stuffed their bags. Where was it taken? Maybe to Kiev, to Minsk. The flea markets. We buried what was left. Dresses, boots, chairs, accordions, sewing machines. Buried it all in pits, which we called 'communal graves'.

I got home. Went to a dance. There was a girl I liked. 'Let's go out together.'

'What for? You're a Chernobyl guy now. Who'd want to marry you?'

I met another girl. We were kissing and cuddling, things were getting serious. 'Let's get married,' I said. And she asked something like: 'You mean you can do that? It's all in working order?'

I wouldn't mind leaving here. That's probably what I'll do. I just worry about my parents.

I have my own memories of it.

Officially, my post there was commander of the security platoon. Something like 'director of the apocalypse zone'. (*He laughs.*) You can write that.

We stopped a truck coming from Pripyat. The town was already evacuated, there were no people left. 'Let's see your documents.' They didn't have any. The back of the truck was covered with a tarpaulin. We lifted it up: there were twenty tea sets, as I recall, a shelving unit, kitchen seating, a TV, carpets and bicycles.

I filed a report.

They brought meat for disposal in the burial sites. The hips were missing from the beef carcasses. The fillet.

I filed a report.

We had a tip-off that a house in an abandoned village was being dismantled. They were numbering and placing the logs on to a tractor with a trailer. We headed straight out to the address given. The raiders were arrested. They were hoping to remove the building and sell it as a dacha. They'd already received advance payment from the future owners.

I filed a report.

Pigs that had gone wild were running about the empty villages. And dogs and cats were waiting by their gates for their owners, keeping watch over the empty houses.

You'd be standing by a communal grave. A cracked stone lists the surnames: Captain Borodin, Senior Lieutenant . . . Long columns, like poetry, with the names of privates. Nettles and burdock . . .

Suddenly, in one of the previously checked vegetable plots, there was a man walking behind a plough. He spotted us. 'Okay, lads, no need to shout. We've already signed our papers. We're leaving in the spring.'

'Then why are you ploughing the plot?'

'Oh, that's an autumn job.'

I understood, but I had to file a report.

Oh, you can bugger off, the lot of you.

My wife took the kid and left, the bitch! But I'm not going to hang myself like Vanya Kotov did. I'm not jumping from the sixth floor. The bitch! When I came back from there with a suitcase full of money – bought a car, gave her a mink coat – she lived with me then, the bitch. Wasn't afraid then. (*He sings.*)

A zillion gamma rays won't zap
The hard-on of a Russian chap.

Good little jingle. It's from back there. Want to hear a joke? (*He launches into it.*) A man comes home from the reactor. His wife asks the doctor, 'What should I do with my husband?' 'Give him a wash, a hug and a decontamination.' The bitch! She's afraid of me. Took the kid. (*Suddenly becoming serious.*) The soldiers were working near the reactor. I took them to and from their shifts. 'Okay, lads, I'm going to count to a hundred. That's it! Off you go!' Like everyone else, I had a personal exposure monitor hung round my neck. After the shift, I had to collect them and hand them over to the KGB. To their secretive First Department. There they took readings, wrote something down on our cards, but the amount of roentgens each person got was a military secret. Bastards! Sons of bitches! Some time goes by, then they tell you: 'Stop! You can't do any more!' All that medical information . . . Even when you were leaving, they wouldn't tell you how much you'd had. Bastards! Sons of bitches! Now they're all fighting for power. Fighting for office, holding elections. Want to hear another joke? After the Chernobyl accident, you can eat whatever you like, but make sure you bury your crap in lead. Ha ha. Life is sweet, sucker, but so short.

How can the doctors treat us now? We didn't bring any documents back. Though I tried. Made requests through the appropriate channels. I got three answers, which I've kept. First answer: the documents were destroyed upon expiry of the three-year statutory storage period. Second answer: the documents were destroyed during the post-perestroika downsizing of the army when units were disbanded. Third answer: the documents were destroyed because they were radioactive. Or maybe they were destroyed so nobody would ever know the truth? We were witnesses. But we're going to die soon. How can we help our doctors? I'd like a certificate right now: how much did I get? What did it amount to? I'd show it to my bitch. Prove to her we can survive under any conditions, still marry and have kids.

Here you go. A clean-up worker's prayer: 'Lord, if it be Your will that I can't do it any more, let it be Your will that I don't still feel randy.' Oh, just go fuck yourselves, all of you!

It started out . . . It all started like a crime novel.

At lunchtime, a phone call came through to the factory: Reservist Private So-and-So to report to the municipal army enlistment office to clear up some details in his documentation. As a matter of urgency. Lots of other men like me turned up at the enlistment office. We were greeted by a captain, who told each of us, 'Tomorrow, you're going to Krasnoe village for reservist training.' The next morning, we all turned up at the enlistment building. They took away our passports and military ID, put us on a bus and took us off to an unknown destination. Not a murmur about reservist training. The officers accompanying us met all our questions with silence. 'Hey, brothers! What if we're off to Chernobyl?' someone guessed. An order: 'Silence! Anyone spreading panic will be court-martialled.' After a bit, an explanation: 'We are under martial law now. No idle chatter! Anyone who abandons the Motherland in her hour of need is a traitor.'

On the first day, we saw the nuclear power plant from a distance. On the second, we were already clearing the rubbish around it. Lugging buckets back and forth. We were shovelling with ordinary spades, sweeping with the kind of brooms street cleaners use, cleaning with scrapers. But those spades were clearly for sand and gravel. Not for litter that had bits of everything in it: plastic sheeting, metal fittings, wood, concrete. As we said, battling the atom with spades! In the twentieth century. The tractors and bulldozers used there were driverless, radio-controlled, and we were following them about and sweeping up the debris. We breathed in that dust. In just one shift we'd change the filter in our radiation masks, popularly known as 'muzzles', up to thirty times. They were awkward things, impractical. The guys often tore them off. It was impossible to breathe in them, especially in the heat, if you were in the sun.

When it was over, we spent another three months in reservist training. Did target practice. They taught us to use a new assault rifle. (*Wryly.*) In case of nuclear war, no doubt. We didn't even get a change of clothing: we were in the same tunics and boots as we wore at the reactor.

Then, of course, they got us to sign some form. A non-disclosure agreement. I kept mum. And if they'd let us speak, who could I have told? Very soon after leaving the army, I became second category disabled. At twenty-two years of age. I was working in a factory. The head of the workshop said, 'Stop falling sick, or we'll make you redundant.' And they did. I went to the director. 'You've got no right. I'm a Chernobyl worker. I saved you all, protected you!' 'It wasn't us who sent you out there.'

At night, I wake up to hear my mum's voice. 'My son, why won't you say anything? You aren't asleep, you're lying there with your eyes open. And your light is on.' I don't say anything. Who's willing to listen to me? Or to talk with me in a way that might encourage me to answer. In my language.

I'm lonely.

I'm not afraid of death any more. Not of death as such.

The mystery, though, is how it will happen. My friend died. He puffed right up, till he was like a barrel. And a neighbour was out there too, he was a crane operator. He went coal black, shrivelled up till he was the size of a child. The mystery is how I'll die. If I could choose, I'd want an ordinary death. Not a Chernobyl death. Just one thing that's certain: with my diagnosis, it won't drag out for long. Sense the moment, and put a bullet in your brain. I was in Afghanistan. Things were easier there. With the bullets . . .

I volunteered for Afghanistan. And for Chernobyl too. I asked to go. We were working in Pripyat. The whole town was surrounded by two rows of barbed wire, like a national border. It had these nice clean high-rise blocks, while the streets were covered with a thick layer of sand, chopped down trees. Looked like shots from a science-fiction movie. We were following orders: 'laundering' the town and replacing the top twenty centimetres of soil with a layer of sand. There were no days off. It was like wartime. I still have a newspaper clipping. About the nuclear operator Leonid Toptunov, he was the one on duty that night at the power plant, pressed the red emergency shutdown button a few minutes before the explosion. It didn't work. He was treated in Moscow. 'To save someone, there has to be a body to start with,' the doctors said, throwing up their hands in despair. He only had one little patch on his back that was still clean, that hadn't been zapped. They buried him in Mitino Cemetery, in a coffin lined with foil. There's one and a half metres of concrete slab layered with lead on top of him. His father visits him, he stands there crying. People walk past and say, 'It's your bloody son blew it all up!' He was just the operator, but he's been buried like an alien from outer space.

It would be better if I'd died in Afghanistan! To tell the truth, those are the thoughts I get racked with. Out there, death was an everyday reality. It was no mystery.

From the helicopter . . .

I was flying close to the ground, observing. There were roe deer, wild boars. All scrawny and sleepy, moving in slow motion. They were feeding on the grass that grew there, drinking the water. They didn't realize they needed to leave too. Along with the humans.

To go out there or not to go? To fly or not? I was a Communist, how could I refuse to fly? Two navigators refused, said they had young wives, hadn't had kids yet. How they were humiliated! That was the end of their careers! There was another court too: the court of manliness. The court of honour! You see, there's a buzz, knowing the other guy couldn't do it, but you can. Now I look at things differently. After my nine operations and two heart attacks, I don't judge anyone any more. I can understand them. They were young lads. But I'd still do it all again. That's a fact. The other guy couldn't do it, but I could. A job for real men!

From the sky, what struck you was the massive amount of equipment: heavy helicopters, medium helicopters. The Mi-24 is a helicopter gunship. What could you do in Chernobyl in a gunship? Or in a Mi-2? The pilots, just young guys, were based in the forest near the reactor, soaking up the roentgens. Those were the orders.

Military orders! But why send such large numbers of people to be irradiated? What for? (*Raising his voice to a shout.*) What they needed were specialists, not human fodder.

From up above, you could see everything. The ruined reactor, the mounds of building debris. And a gigantic number of tiny human figures. There was a West German crane, but it was dead; it had moved a little along the roof and then died. The robots were dying. Our Soviet robots, created by our Academician Lukachev for exploring Mars. And the Japanese robot that looked like a human. But it was clear their insides had been fried by the high doses of radiation. The little soldiers were running around in their rubber suits and rubber gloves. They looked so small, seen from the sky.

I fixed it all in my memory. Thought I'd tell my son. But when I got back: 'Daddy, what did you see?' 'A war.' I had no other words for it.

2

The Crown of Creation

Monologue on the old prophecies

My daughter. She's different from other kids. She'll grow up and ask me, 'Why am I different?'

When she was born . . . It wasn't a child, but a little living sack, stitched up on all sides, without a single slit, only the eyes were open. Her medical record said: 'Girl born with complex multiple pathologies: anal aplasia, vaginal aplasia and left renal aplasia.' That's what they call it in scientific language, but in plain words: she has no private parts, no bum, and just the one kidney. On day two, the second day of her life, I took her to be operated on. She opened her eyes like she was smiling, and at first I thought she wanted to cry. Oh my God, she smiled! Other babies like her don't live, they die straight away. What kept her from dying was my love for her. She's had four operations in four years. This is the only child in Belarus to survive with such complex pathologies. I love her so much. (*She pauses.*) I can't give birth ever again. I wouldn't dare. I got back from the maternity ward: my husband would kiss me at night, and I'd tremble all over – but we can't. It would be a sin. I'd be scared to. I overheard the doctors saying, 'That girl wasn't born in a caul, she was born in a shell. If they showed her on TV, not one mother would ever give birth again.' That's what they said about our little girl. How could we love each other again after that?

I went to church, told the priest. He said we needed to pray for our sins. But none of our family has ever murdered anyone. What is it I'm guilty of? At first, they wanted to evacuate our village, but then they crossed us off the list: the authorities didn't have the money. And that's when I fell in love. Got married. I didn't know that we couldn't love each other here. Many years ago, my grandmother read in the Bible that there would come a time on earth when everything would be lush, everything would blossom and bear fruit, the rivers would be teeming with fish, the forests full of animals, but man wouldn't be able to use any of it. And we wouldn't be able to produce our own kind, prolong our immortality. I listened to those old prophecies like they were scary tales. Didn't believe them. Please tell everyone about my little girl. Write about her. At four years old, she can sing and dance, knows poems by heart. She has normal intelligence, she's no different from other kids, it's just that she plays different games. She won't play at shopping or schools, she'll play at hospitals with her dolls: give them jabs, take their temperature, put them on drips. If the doll dies, she covers it up with a white sheet. For four years, we've been living with her in the hospital; she can't be left on her own there, and she doesn't know that people are actually meant to live at home. When I take her home for a month or two, she asks, 'Are we going back to the hospital soon?' Her friends are there, it's where they live and grow up. They've made her a bum, they're forming the vagina. After the last operation, she completely stopped passing water, they didn't manage to insert the catheter – she'll need more operations. But they're recommending that we continue with the surgery abroad. And where are we going to find tens of thousands of dollars, when my husband is on 120 dollars a month? One professor secretly suggested, 'With those pathologies, your child should be of real interest to science. Write to some foreign clinics. They ought to be interested.' And so I'm writing . . . (*Fighting back the tears.*) I'm writing that every half hour she has to have her urine pressed out by hand; the urine comes out through these little holes around the vagina. If it isn't done, her single kidney will fail. Where else in the world can you find a child who needs to have their urine pressed out every half hour by hand? And how much longer can we take this? (*She cries.*) I don't allow myself to cry. I mustn't cry. I'm trying all the doors, writing letters. Please take my little girl, even if it's for experiments, for research. I'd agree to her becoming a guinea pig, a lab rat, if only they could save her. (*Crying.*) I've written dozens of letters. Dear God!

So far, she hasn't understood, but one day she'll ask us why she's different from others. Why no man will be able to love her. Why she won't be able to have a baby. Why she'll never be able to do what the butterflies do, what the birds do, what everyone but her can do. I wanted . . . I needed to prove . . . that . . . I wanted to get documents . . . so she'd grow up and find out it wasn't me and my husband to blame. It wasn't our love. (*Fighting back the tears again.*) For four years, I've been at war. With the doctors and officials. Trying to get an audience with people in high places. It took me four years to get a medical certificate making the link between low-level ionizing radiation and her dreadful pathologies. For four years, they've refused to, kept repeating, 'Your little girl is classed as "disabled from childhood".' What do they mean, 'disabled from childhood'? She's disabled from Chernobyl. I've studied my family tree: there hasn't been a case like this, they've all lived till eighty or ninety, my grandfather lived to ninety-four. The doctors came up with excuses. 'We need to follow protocol. We still have to treat it as due to natural causes. Only in twenty or thirty years' time, when we've built up a data bank, can we link these diseases to low-level ionizing radiation. And to what we're eating and drinking on our land. But, for the moment, too little is known by medicine and science.' Well, I can't wait twenty or thirty years. That's half a lifetime! I wanted to take them to court, take the state itself to court. They've called me crazy, laughed at me. They said children like that were born even in ancient Greece, ancient China. One official shouted, 'You're after Chernobyl benefits! Chernobyl money!' I'm amazed I didn't faint in his office, die of a heart attack. But I mustn't allow that.

There was one thing they just couldn't, and wouldn't, understand: I had to know that I and my husband weren't to blame. That it wasn't our love. (*She faces the window and cries quietly.*)

This is a little girl growing up. No matter what, she's a little girl. I don't want you using our surname. Even the neighbours on our landing don't know the whole story. I'll put her in a dress, plait her hair. 'Your little Katya is so pretty,' they tell me. I look at pregnant women in such a funny way now. Like I'm seeing them from a distance, from round a corner. I don't look so much as peep at them. Inside, there's a mess of emotions: surprise and horror, envy and joy, there's even some vindictiveness. Once I caught myself looking with the same emotion at our neighbour's pregnant dog. At a mother stork in her nest . . .

My little girl . . .

Larisa Z., mother

Monologue on a moonscape

All of a sudden, I've started wondering whether it's better to remember or forget.

I've asked my friends. Some have forgotten, others don't want to remember, because there's nothing we can change, we can't even move out of this place. Can't even do that.

Here are my memories. In the days straight after the accident, all the books on radiation – on Hiroshima and Nagasaki, even on X-rays – vanished from the library. There was a rumour going around that it was on orders from the authorities, to keep people from panicking. For our own peace of mind. There was even a joke that, if Chernobyl had exploded on Papua New Guinea, everyone but the Papuans would be shaking with fear. There was no medical advice, no information. Whoever could, got their hands on potassium iodide tablets (the pharmacies in our town didn't stock it, you had to know people in the right places). Some people took a handful of the pills and swallowed them down with a glass of pure alcohol. The paramedics saved their lives.

The first foreign journalists came; the first film crew. They were all in plastic boiler suits, helmets, rubber overshoes, gloves – even the camera was in a special case. And accompanying them was one of our Soviet girls, interpreting. In a summer dress and sandals.

People believed every word in print, although no one was printing the truth. They weren't speaking the truth. While it's true they were hiding things, at the same time there was plenty they just didn't understand. From the general secretary of the Party down to the road sweepers. Later came the signs, everyone was following them closely. So long as the town or village has sparrows and pigeons, it's safe for people to live there. If the bees are busy, it's still clean. I was in a taxi, and the driver was puzzling over why the birds were acting as though they were blind, dropping down on his windscreen, crashing into it. Like they'd gone daft and dozy. They seemed suicidal. After his shift, to blank it all out, he used to sit drinking with his friends.

I have memories of coming back from a trip for work. On both sides of the road there was a genuine moonscape: fields covered in white dolomite, stretching out to the horizon. The top layer of contaminated soil had been removed and buried, with this dolomite sand poured in its place. As if it wasn't earth; like we weren't on earth. I was haunted by that vision for a long time and tried writing a story. I imagined what it would look like in a hundred years' time: a creature, maybe human, maybe something else, hopping on four legs, jerking its long hind legs back with its knees, who sees by night with its third eye, and with its one ear on the top of its head it can hear even an ant running. Ants are all that's left, everything else on earth and in the sky has died out.

I sent the story off to a magazine. They wrote back saying it wasn't a work of literature, it was a description of horror. Of course, I didn't have the talent. But on this occasion, I suspected another reason was at play. I began wondering why so little has been written on Chernobyl. Our writers keep on writing about the war, about Stalin's camps, but they're silent on Chernobyl. There are almost no books on it. Do you think that's just a coincidence? It's an episode still outside our culture. Too traumatic for our culture. And our only answer is silence. We just close our eyes, like little children, and think we can hide. Something from the future is peeking out and it's just too big for our minds. Too huge for us to handle. If you get chatting with somebody, he'll start telling his story, and he'll thank you for listening to him. You may not have understood, but at least you listened. Because he didn't understand it either. Just like you. I've gone off reading science fiction.

So which is better: remembering or forgetting?

Yevgeny Alexandrovich Brovkin, lecturer at Gomel State University

Monologue of a witness who had toothache when he saw Christ fall and cry out

At the time, I had other things on my mind. It might sound strange to you. This was when I was getting a divorce from my wife.

They suddenly turned up, handed me a call-up notice and said there was a van waiting downstairs. A special 'meat wagon'. Just like in 1937. They came for us by night; took us warm from our beds. Then later this set-up stopped working: the wives wouldn't open the door, or they'd fib about their husbands being off on a business trip, away on holiday, visiting their parents in the country-side. They tried to hand over the call-up notices, but the wives wouldn't take them. So they started seizing people at work, in the streets, during their lunch break in the factory canteens. Just like in 1937. And at the time, I was on the verge of going crazy. My wife had been unfaithful, everything else seemed like no big deal. I got into the meat wagon. Two men led me to it – in plain clothes, though they carried themselves like soldiers – they walked on either side of me, obviously worried I'd run for it. When I got in the van, for some reason I thought of the American astronauts who flew to the moon, and one of them later became a priest, while the other, I think he went mad? I read about their visions. They were meant to have seen these remains of cities, with human traces. These press clippings flashed through my mind: our nuclear power plants are completely safe, you could build one right in Red Square. Next to the Kremlin. Safer than a samovar. They were like the stars, and we would sprinkle them all over the land. But my wife had left me. That was all I was capable of thinking about. I tried killing myself a few times, swallowed pills, wishing never to wake up. We were at the same kindergarten, went to the same school. Studied at the same college. (*He lights a cigarette and stops speaking.*)

I warned you. Nothing heroic here for the writer's pen. Had thoughts about how this wasn't wartime, why I should put myself at risk while someone else was sleeping with my wife. Why me and not him? To be honest, I didn't see any heroes there. I saw madmen who couldn't give a damn about their own lives; saw enough bravado, but there was no need for it. I also have the certificates of merit and commendation. But that's because I wasn't afraid to die; couldn't give a damn! Even saw it as a way out. They would have buried me with honours. And at public expense.

Out there, you were thrown into this dreamlike world where the End of Time met the Stone Age. On the inside, it all still felt raw and intense. We were living in the forest, in tents. Twenty kilometres from the reactor. We were being 'partisans'. 'Partisans' means the reservists called up for training. Our ages ranged from twenty-five to forty, many of us were university educated or vocationally trained. Myself, I'm a history teacher. Instead of assault rifles they gave us spades. We dug up refuse dumps and vegetable plots. The women in the villages looked at us and crossed themselves. We were in gloves, respirators and camouflage suits. The sun was baking down. We'd turn up at someone's vegetable plot like some kind of devils, looking like aliens. They didn't understand why we were digging up their beds, ripping out their garlic and cabbage, when it was just your ordinary garlic, ordinary cabbage. The old women crossed themselves and wailed: 'What is it, soldiers, the end of the world?'

Inside a house, they'd have the stove lit and pork fat frying. You'd bring the dosimeter up close: that stove was a mini nuclear reactor. 'Sit down, lads, join us at the table,' they'd say. They were welcoming us into their home. We turned them down. They said, 'We'll find you a hundred mils of vodka. Sit down. Tell us about it all.' What could we tell them? At the reactor itself, the firemen had been trampling soft fuel, which was glowing, but they hadn't known what it was. So how were we going to know anything?

Our squad would set out. We had just the one dosimeter for all of us. But you'd get a different reading in each place: one guy would be working where it was two roentgens, another where it was ten. We felt as powerless as prisoners, and at the same time there was fear. And mystery. But I didn't experience any fear; I was looking at everything from the outside.

A group of scientists flew in by helicopter. They were wearing rubber suits, high boots, safety glasses. Proper astronauts. An old woman went up to one of them: 'Who are you?' 'I'm a scientist.' 'Oh, so you're a scientist! Look at him, in that get-up. You've camouflaged yourself. And what about us?' She chased him with her stick. The thought occurred to me a few times that some day they'll start hunting down scientists the way they caught and drowned doctors in the Middle Ages, burned them at the stake.

I saw a man watching as they buried his home. (*He gets up and walks over to the window.*) There was just a freshly dug grave left. A large rectangle. They buried the well, the orchard. (*He falls silent.*) We were burying earth. We would carve it out and coil it into big rolls. I warned you. There was nothing heroic.

We used to get back late in the evening because we were working twelve hours a day, seven days a week. The only time off was at night. One time, we were driving in an armoured personnel carrier. Somebody was walking through the empty village. We got closer: it was a young fellow carrying a carpet on his shoulders. Not far off was a Lada. We stopped the vehicle. The car boot was chock-full of TVs and telephones with the cords cut off. We turned around and rammed the APC into the Lada, which crumpled like an accordion, as if it was a tin can. No one said a word.

We were burying the forest. We sawed up trees into one-and-a-half-metre lengths, packed them in plastic and dumped them in the burial site. At night, I couldn't sleep. I'd shut my eyes, and there'd be something black wriggling and twisting. Like it was alive; living rolls of earth. Full of beetles, spiders and worms. I didn't know what any of them were, didn't know their names. They were just beetles, spiders. Ants. But there were little and large ones, some were yellow, some were black. They were all different colours. There was a poet who wrote that the animals are a separate nation.

I was killing them by the tens, hundreds, thousands, without even knowing their names. I was smashing their homes, their secrets. Burying and burying them.

Leonid Andreyev, a writer I particularly love, wrote a parable about Lazarus, who had seen beyond what's allowed. He turned into an outsider, could never again be like everyone else, even though he'd been raised from the dead by Christ.

Maybe that's enough, now? You're curious, I get that. They're always curious, the people who weren't there. There's a Chernobyl in Minsk and another one in the actual Zone. Somewhere in Europe you'll find a third one. In the Zone itself I was struck by the indifference with which people talked about the disaster. In one dead village, we met an old man. He was living all alone. We asked him, 'Aren't you afraid?' And he answered, 'Of what?' You can't be frightened the whole time, a person can't do that; some time goes by, and ordinary human life starts up again. Everyday human life. Our men were drinking vodka, playing cards, chatting up women. They talked a lot about money, but they weren't working for the money. Very few people were only after the money. They worked because they had to. They were told to, and didn't ask questions. They dreamed of promotions, they schemed and stole things. They hoped for the promised benefits: jumping the queue for a flat so they could move out of their factory barracks, getting their kids a kindergarten place, buying a car. One of the guys wimped out, got all frightened of leaving the tent, slept in a homemade rubber suit. The coward! They expelled him from the Party. He used to shout, 'I want to live!' Everything was all mixed up. I met women out there who'd volunteered and were raring to go. The authorities had turned them down, explained they needed drivers, mechanics and firemen, but the women came anyway. Everything was all mixed up. There were thousands of volunteers. You had the volunteer student teams, and then the special meat wagons stalking reservists by night. The charity collections, the money transfers to the relief fund, the hundreds of people selflessly donating their blood and bone marrow. And at the same time, there was nothing you couldn't buy for a bottle of vodka. You could buy home leave, a certificate of merit. One collective-farm chairman would bring a crate of vodka for the radiation monitoring technicians to keep his village off the evacuation list, while another brought a crate of vodka so his collective farm would be resettled. The fellow had been promised a two-bedroom apartment in Minsk. Nobody was checking the radiation readings given. It was the usual Russian chaos. That's how our life is. Things got struck off the inventories and sold. It was a case of: sure, it's sickening, but to hell with you all!

Students were sent to join us. They pulled up orache from the fields and raked the hay. There were some really young couples, husband and wife. They even walked about holding hands. It was unbearable to look at. But all these places were so beautiful! Such grandeur. The horror was made all the worse by that beauty. And human beings just had to get out of there. They had to flee like villains, or criminals.

Every day, they brought the papers. I read just the headlines: 'Chernobyl, Site of Heroic Deeds', 'Reactor Is Vanquished', 'Life Goes On', We had political officers. They put on political sessions, told us we had to emerge victorious. Over what? The atom? Physics? The cosmos? Our nation treats victory not as an event but a process. Life is a battle. That's where our great love of floods, fires and earthquakes comes from. We need a stage for our 'displays of courage and heroism'. Somewhere to hoist the flag. The political officer read us news items on the 'high level of political awareness and efficient organization', on how, within a few days of the accident, the red flag was flying over Reactor No. 4. There it proudly fluttered, until a few months later it was ravaged by the tremendous radiation. So they raised a new flag. And another. The old one was kept as a souvenir. They ripped it into shreds and shoved it under their jackets next to their hearts. Then they took the rags back home, showed them off proudly to their children. They preserved them. Heroic lunacy! But I was just the same, no better. I tried to imagine the soldiers climbing on to the roof. Sheer suicide. But they were brimming with emotions. First, their sense of duty and, second, passion for the Motherland. You'll tell me it was Soviet paganism? But the thing is, if they'd handed the flag to me, I would have climbed up there too. Why? I can't really answer that. Of course, the fact that I didn't care if I died played a role. My wife didn't even send me letters. In six months I didn't get one. (*He stops.*)

Want to hear a joke? A prisoner breaks out of jail. He hides in the thirty-kilometre zone. They catch him and take him to the radiation monitoring technicians. He's glowing so much they can't throw him back in jail, put him in hospital or leave him anywhere near people. (*He laughs.*) We loved jokes out there. Black humour.

I arrived there when the birds were sitting in their nests, and left when the apples were lying on the snow. There was a lot we didn't manage to bury. We buried earth in the earth. Along with the beetles, spiders and maggots, that whole separate nation. We buried a world. That was the deepest impression I came away with. Those creatures.

I haven't told you anything, really. Just some fragments. Leonid Andreyev had another story. There was a man who lived in Jerusalem, and Christ was led past his house. He saw and heard everything, but at the time he was suffering from toothache. Right before his eyes, Christ stumbled and fell as He carried the Cross, and He began crying out. The man saw everything, but he had toothache and didn't run out into the street. A couple of days later, when his tooth had stopped hurting, he was told how Christ had been raised from the dead. That's when he realized, 'I could have been a witness, but I had toothache.'

Is that how it always has to be? People will never prove equal to a great event; they're always out of their depth. My father defended Moscow in '42. The fact that he was taking part in history only sunk in decades later. He realized it from the books and films. All he told us was, 'I sat firing from a trench. Got buried by a blast. The medics dragged me out half-dead.' That was it.

And at the time, my wife had left me.

Arkady Filin, clean-up worker

*Three monologues on the 'walking ashes' and the
'talking dust'*

> *Viktor Iosifovich Verzhikovsky, chairman of the
> Khoyniki Hunters and Anglers Society, and two hunters, Andrey
> and Vladimir, who declined to give their surnames.*

I was a child the first time I killed a fox. The second time it was a cow elk. I vowed never to kill one again. They have such expressive eyes.

It's us humans who understand things, the animals just live. And the birds too.

In the autumn the female roe deer is very sensitive. If, on top of that, she is downwind of you, you've had it. She won't let you near. But a fox is a sly old thing.

There's one guy kicking about here. Once he's had a few, he'll start making speeches to everyone. He studied philosophy at university, then did time in prison. When you meet people in the Zone, they never tell you the truth about themselves. Or rarely. But this one has his head screwed on right. 'Chernobyl,' he'll say, 'happened so as to give us philosophers.' He called animals 'walking ashes', and people were 'talking dust'. 'Talking dust' because dust we are, and to dust we shall return.

The Zone pulls you in. Like a magnet, I'm telling you. Saints preserve us! Once you've been there . . . Your heart is drawn to it.

I read this book. There were saints who talked with the birds and the beasts. And we think they can't understand humans.

Well, lads, we need to start at the beginning.

Go ahead, chairman, and meantime we'll light up.

*

So then, here's how it went. I got called to the District Executive Committee. 'Look here, chief hunter: we have a lot of pets left behind in the Zone, cats and dogs, they need to be shot in the interests of disease prevention. Go to work!' The next day, I called everyone over, all the hunters. I announced that blah-di-blah. Nobody wanted to go, because they weren't issuing any protective equipment. I tried civil defence – they didn't have anything. Not a single respirator. I ended up going to the cement plant and getting masks there. Made with a very fine weave, for cement dust. But they didn't have respirators for us.

We ran into soldiers there. They were in masks and gloves, in armoured personnel carriers, while we were in our shirts, with gauze masks tied round our noses. We went back home in those shirts and boots. Back to our families.

I cobbled together two teams. Some volunteers signed up too. Two teams, twenty people in each. Each team was allocated a vet and someone from health and disease control. We also had a tractor with an excavator bucket and a tipper truck. It was wrong they didn't give us protective equipment. They didn't spare a thought for the people.

But they did give us bonuses: thirty roubles each. A bottle of vodka in those days cost you three roubles. We decontaminated ourselves. Recipes sprang up from nowhere: put one spoonful of goose droppings in a bottle of vodka, let it stand for two days and drink it. To keep our, you know, family jewels safe and sound. Remember all those folk ditties? Heaps of them.

> A Trabant is a heap of junk.
> A man from Kiev has no spunk.
> If you want to shine in bed,
> Better wrap your balls in lead.

Ha ha.

*

We drove around the Zone for two months. In our district, half the villages had been evacuated. There were dozens of them: Babchin, Tulgovichi . . . The first time we went, dogs were running about near their houses. Guarding them, waiting for people to come back. They were excited to see us, came running to a human voice. They welcomed us. We shot them in the houses, the barns, the vegetable plots. Then we dragged them out and loaded them on to the tipper trucks. Not pleasant, of course. They couldn't understand why we were killing them. They were easy to kill. These are pets: they don't fear guns, don't fear man, come running to a human voice.

There was a tortoise crawling along. Good Lord! Past an empty house. The apartments had aquariums in them, with fish.

We didn't kill the tortoises. You can run the front wheel of a jeep over a tortoise shell and it will take it, won't crack. Only if you were drunk, of course, running over it. In the yards, there were cages wide open, rabbits running about. The coypus were locked up, we let them out; if there was any water nearby, a lake or river, they swam away. Everything had been left in a hurry. Thought they'd soon be back. After all, how had it gone? An evacuation order 'for three days'. Women bawling, children crying, cows bellowing. The little ones were tricked: 'We're going to the circus.' People thought they would be back. Nobody said this was forever. Ah, saints preserve us! I'm telling you, it was a war zone. Cats looking into people's eyes, their dogs howling, trying to get on the buses. Mongrels, Alsatians. The soldiers were pushing them out again, kicking them. They ran after the buses for ages. Evacuation . . . God forbid we ever have another!

So then, here's how it works. The Japanese had their Hiroshima, and now look at them: ahead of the pack. Number one in the whole world. So then . . .

We were given the chance to go shooting and, what's more, with live, running prey. The thrill of the chase. We'd down some drinks and off we'd go. At my workplace, they counted it as a day's work, paying me for it. They could, of course, have topped it up, considering the nature of the job. The bonus was thirty roubles, but the money wasn't worth as much as under the Communists. Everything had changed already.

Here's how it worked. At first, the houses were all sealed up. We didn't rip off the seals. Through the window, you'd see a cat sitting inside. How could you get at it? We left them alone. Later, the looters started breaking in – they kicked the doors in, smashed the windows and gutted the place. The first things to go would be tape recorders and TVs. All the furs. Then, later, they stripped everything clean. Just some aluminium spoons left lying on the floor. And the surviving dogs moved into the houses. If you went in, they'd go for you. They'd stopped trusting humans. I went in one place, and there was a bitch lying in the middle of the room with her puppies. Did I feel sorry for them? Not pleasant, of course. I thought about it. We were basically acting like a death squad. Like in wartime, following the same plan. A military operation. We too would arrive, seal off a village, and as soon as they heard the first shot, the dogs would scram. Run for the forest. The cats were smarter, better at hiding. A kitten hid in a clay pot. I shook it out. Pulled them out from under the stove. Not a pleasant feeling. You go in the house and the cat darts past your boots like a bullet, you run after it with your rifle. They're all skinny and dirty. Patchy fur. At first, there were heaps of eggs, lots of hens had been left behind. The dogs and cats were eating the eggs, and when those ran out, they ate the hens. The foxes were eating the hens too. By now, foxes were living in the village alongside the dogs. So, when those ran out of hens, the dogs began eating the cats. Sometimes we came across pigs in the barns. We let them out. In the cellars, there were all sorts of jars: gherkins, tomatoes. We'd open some and throw them in the trough. We didn't kill the pigs.

We found an old woman. She locked herself in her cottage: had five cats and three dogs. 'Don't hurt the dog: once she was human too.'

She wouldn't hand them over. Cursed us. We took them by force, but left her one cat and one dog. She called us names: 'You bandits! You jailers!'

Ha ha.

> By the hill there ploughs a tractor,
> Up the hill burns a reactor.
> Had not the Swedes made us aware,
> We would still be ploughing there.

Ha ha.

Empty villages. Just the stoves left. It's what the Nazis did to Khatyn all over again! There's an old man and old woman living there, like something out of a fairy tale. They aren't afraid, but anyone else would have lost their minds! By night, they burn old tree stumps. The wolves are frightened of the fire.

So then, here's how it was. The smells. I just couldn't work out where that smell in the village was coming from. Six kilometres from the reactor. The village of Masaly, it smelled like an X-ray room. Reeked of iodine, some kind of acid. But they say radiation has no smell: I don't know. We had to shoot at point-blank. So this bitch was in the middle of the room with her puppies. She went for me – I put a bullet in her. The puppies were licking my arms, being all sweet and playful. We had to shoot at point-blank. Saints preserve us! There was this one dog, a little black poodle. I still feel sorry for it. We heaped the tipper full of them. Taking them to the burial site. To tell the truth, it was just a plain old deep pit, though you're meant to dig it taking care not to reach the ground water, and line the bottom with plastic. You're meant to find some spot fairly high up, but you know how it is. The rules were broken all the time: we had no plastic, and we didn't spend long looking for the right spot. If you wound them rather than killing them, they'll squeal and cry. They were tipping them out of the truck into the pit, and this little poodle began scrabbling about. It climbed out. Nobody had any cartridges left. Had nothing to finish it off with, not a single cartridge. They shoved it back into the pit and covered them all up with earth. Still feel sorry for it.

There weren't nearly as many cats, though. Maybe they'd left in search of people? Or were they hiding? A poodle is an indoor, pampered little thing.

It's better to kill from a distance, so your eyes don't meet.

Make sure you learn to shoot straight, so you won't have to finish them off.

It's us humans who understand things, while they just live. 'Walking ashes.'

The horses . . . They were being taken to slaughter. They were crying.

And I'll add this. Every living creature has a soul. From boyhood, my father trained me to hunt. When a doe is lying there wounded, she wants to be pitied, but you're trying to finish her off. In her last moment, she has a completely conscious, almost human look in her eyes. She hates you. Or she's pleading: 'Let me live! I want to live too!'

Learn to shoot straight! I'm telling you, having to finish them off is a lot more unpleasant than a clean kill. Hunting is a sport, a form of sport. Why does nobody go on at the anglers, while the hunters get all the abuse. It's not fair!

Hunting and war have always been the most important pursuits of men. Ever since time began.

I couldn't admit to my son, my little boy, where I had been, what I was doing. To this day, he thinks his daddy was defending someone. Standing guard somewhere! On the television, they were showing all the military hardware, the soldiers. A lot of soldiers. My son asked, 'Daddy, were you a soldier there?'

A TV cameraman came out with us. Remember? He had a movie camera. It made him cry. He was a bloke, but he cried. He was hoping to see a three-headed boar or something.

Ha ha. Fox spies Little Round Bun rolling through the forest. 'Little Round Bun, where are you rolling to?' 'I'm not Little Round Bun, I'm a Chernobyl hedgehog.' Ha ha. And as they say: 'Atoms for peace: every home should have one!'

Man dies like an animal, I'm telling you. I've seen it, many times. In Afghanistan. I was wounded in the stomach, lay there in the sun. The heat was unbearable. Oh, for a sip of water! 'So,' I think, 'I'm going to die here like a dog.' I'm telling you, our blood flows just the same as theirs. And pain is the same.

The policeman we had with us, you know . . . went cuckoo. Ended up in hospital. He was always feeling sorry for the Siamese cats, saying how much they cost at the market, and how beautiful they were. Wasn't quite right in the head.

A cow was walking with her calf. We didn't shoot them. Nor did we shoot the horses. They were frightened of the wolves, but not of human beings. But horses are better at defending themselves. It was the cows that the wolves got first. That's the law of the jungle.

They transported cattle from Belarus and sold them to Russia. The heifers had leukaemia, so they flogged them off cheap.

More than anything, I felt sorry for the old men. They'd come up to our vehicles. 'Look in on my house, lad.' They'd put the keys in your hand. 'You can bring me my suit, and my hat.' They'd give you some cash. 'See how my dog's doing.' The dog had been shot and the house looted. But the old guys would never be going back, anyway. How could you break it to them? I wouldn't take their keys, didn't want to cheat them. Others did. 'Where's your moonshine kept? Where do you hide it?' The old grandad would tell them and they'd find whole churns of the stuff, big milk churns.

They asked us to shoot a wild boar for a wedding. Placed an order! The liver was like jelly in your hands, but they wanted it anyway. For weddings and christenings.

We also go shooting for science. Once a quarter, we catch two rabbits, two foxes and two deer. Contaminated, all of them. But we hunt and eat them anyway. At first, we were worried, but now we're used to it. We've got to eat something, can't all move to the moon or some other planet.

One chap bought a fox-fur hat at the market: he went bald. An Armenian bought a dirt-cheap assault rifle from a burial site: he died. We used to spook each other out.

My heart and head were unaffected there. Moggies and doggies. Saints preserve us! I shot them. That was my job.

I was talking to this driver who was bringing out houses from there. The Zone is being stripped down and sold. Though they aren't schools, houses and kindergartens any more: they're inventorized items for decontamination. They're being carted away! We met him maybe in the bathhouse or near a beer stand, I don't remember exactly. What he said was, they drive out there in KamAZ trucks, dismantle a house in three hours, and near the town they are stopped by buyers. Those dacha parts, they sell like hot cakes. The Zone has been sold off for dachas.

There are predators among us. Hunter predators. Other people just enjoy walking through the forest, hunting small game and birds.

*

I'm telling you. So many people suffered, and nobody has answered for it. They flung the management of the atomic power station in prison, and that was all. Under that system, it was very hard to say who was to blame. When you got orders from above, what could you do? Only one thing: carry them out. They were testing something there. I read in the papers that the military were stockpiling plutonium. For atomic bombs. That's why it all came crashing down. To put it crudely, the question is: why Chernobyl? Why did it happen to us, and not to the French or the Germans?

It's lodged in my memory. That's how it was. A pity nobody had a single cartridge left, nothing to shoot it with. That poodle. There were twenty of us. Not a single cartridge by the end of the day.

Monologue on how we can't live without Tolstoy and Chekhov

What do I pray for? If you were to ask me what I pray for . . . I say my prayers at home, not in church. In the morning or evening, when everyone is asleep.

I want to feel love. I do feel love! And I pray for that love. But should I . . . (*She trails off mid-sentence. I can see she is reluctant to go on.*) . . . remember? Maybe, at any rate I need to shed it all, put it aside. I've never read about this in any books, never seen anything like it in films. The films just showed war. My grandparents used to say they had no childhood, just war. Their childhood was the war, and mine was Chernobyl. That's where I'm from. Look, you're writing this, but not one book has ever helped me, they've never explained it. Nor did the theatre or films. I've worked it all out for myself, without their help. We're living through all this ourselves, and we don't know what to do. I can't get my head round it. My mother was completely at sea: she teaches language and literature in a school. Always taught me to live life by books. Then suddenly there are no books to guide you. Mum was lost, she couldn't live without her books. Without Tolstoy and Chekhov.

Memories? I both want to remember and don't. (*She appears to be listening to an inner voice, or debating with herself.*) If the scientists don't know anything, if the writers don't either, then we can help them with our own lives and deaths. That's what my mother reckons. But I'd rather not think about it. I'd like to be happy. Why can't I be happy?

We lived in Pripyat, near the atomic power station. It's where I was born and raised. In a big prefab apartment block, up on the fourth floor. The windows looked out on the power plant. On 26 April . . . Many people told me later they definitely heard the explosion. I don't know; in our family, no one noticed it. I woke up in the morning as usual, to go to school. I heard a rumbling. Through the window, I saw a helicopter hovering over the roof of our building. 'Blimey!' I thought. 'This'll give me something to tell the kids about at school!' How was I to know we only had two days left? Of our old life. There were two more days, our last days in the town. The town is gone now. What's left isn't our town any more. I remember our neighbour sitting on the balcony with binoculars, watching the fire. It was about three kilometres away as the crow flies. And in the afternoon, I cycled over with the girls and boys to the power station. The kids who didn't have bikes were jealous of us. Nobody told us off, nobody! Not our parents, not our teachers. By lunchtime, no anglers were left on the river: they came home brown. Even a month in Sochi wouldn't get you that colour. A nuclear tan! The smoke hanging over the power plant wasn't black or yellow, it was blue, it had this blue tinge. But nobody told us off. Must've been our upbringing. We'd grown used to the idea that danger could only come from war, with explosions left, right and centre. But here we had an ordinary fire being put out by ordinary firemen. The boys were cracking jokes: 'Line up in long ranks towards the cemetery. Whoever's tallest will be the first to die.' I was small, myself. I don't recall any fear, but I remember a lot of things that were strange, unusual. A friend told me how she and her mum buried money and gold jewellery in the yard at night, and they worried about forgetting the spot. My grandma had been given a Tula samovar for her retirement, and for some reason that samovar and Grandpa's medals were what bothered her most. And her old Singer sewing machine. She was wondering where to hide them. Soon we were being evacuated. It was a word Dad brought home from work: 'We're leaving with the evacuation.' Just like in war novels. When we were on board the bus, Dad remembered something he'd forgotten to bring. He ran home. He returned with two of his new shirts, still on their hangers. It was odd, not like him. Inside the bus, we all sat in silence, staring out the window. The soldiers looked unearthly, they were walking about the streets in white snow suits and face masks. People would go up to them and ask, 'What's going to happen to us?' 'Why ask us?' they'd say stroppily. 'Ask those guys over there in the white Volgas, they're in command.'

We were leaving. The sky was as blue as can be. Where were we going? People had Easter cakes and painted eggs in their bags and string shoppers. If this was war, then from the books I'd imagined it differently. With explosions left, right and centre, and air raids. We moved slowly, held up by cattle. Cows and horses were being herded along the roads. There was a smell of dust and milk. The drivers swore and shouted at the herdsmen: 'What the Devil are you doing, driving your herd down the road, you mother-effers! You're stirring up radioactive dust! Couldn't you have gone through the field or the meadow?' The herdsmen responded in kind, making excuses, saying it would be a shame to trample the young wheat and grass. No one believed we wouldn't be returning. For people not to go back home – that had never happened before. I felt a bit dizzy and my throat was tickling. The older women didn't cry, but the young ones did. My mother was crying.

We made our way to Minsk. We bought train tickets from the conductor for three times the price. She brought in tea for everyone, but to us she said, 'Use your own mugs and glasses.' It took a while for us to cotton on. First we wondered if they were short of glasses. No, that wasn't it! They were frightened of us. We'd get asked, 'Where are you from?' 'Chernobyl.' And they would sidle away from our compartment, not let their children run past us. We arrived in Minsk, got to my mother's friend's place. My mum to this day feels ashamed that we piled into someone else's apartment in our 'dirty' clothes and shoes that night. But they welcomed us and fed us. Felt sorry for us. Although their neighbours popped round: 'You've got guests? Where are they from?' 'Chernobyl.' And they sidled away too.

A month later, they allowed my parents to go back and check on the apartment. They brought back a warm blanket, my autumn coat and Chekhov's Collected Letters, my mum's favourite. I think it was seven volumes. Grandma, though . . . Our grandma couldn't understand why Mum hadn't picked up a couple of jars of the strawberry jam that I loved. It was sealed up in jars, with the lids closed. Metal lids. On the blanket we found a 'spot'. Mum tried washing it, vacuuming it, but nothing helped. We took it to the dry cleaners. But it carried on glowing, this spot, until we cut it out with scissors. Here were our old familiar things: a blanket, a coat. But I couldn't sleep under the blanket any more, couldn't wear the coat. We didn't have the money to buy a new coat for me, but I couldn't use that one. I hated those things, that coat! It wasn't that I was afraid, you know, I really did hate them. Those things could kill me! And they could kill my mother! That feeling of hatred . . . I couldn't get my head round any of it. They were talking everywhere about the disaster: at home, in school, on the bus, in the streets. They were comparing it with Hiroshima, but no one believed that. How can you believe anything if it's baffling? No matter how hard you strain and try, it doesn't get any clearer. I remember when we drove out of our town – the sky was as blue as can be.

Grandma . . . She didn't take to the new place. She was home-sick. Just before she died, she told us: 'I want sorrel!' For years they wouldn't let us eat sorrel, it's worse than anything for soaking up radiation. We took her back to be buried in her native village of Dubrovniki. By then, it was inside the Zone, fenced off by barbed wire. There were soldiers with Kalashnikovs standing guard. They only let adults beyond the barbed wire: Mum and Dad and our relatives. But I wasn't allowed in: 'No children.' I realized I would never be able to visit my grandmother. It sank in. Who ever heard of anything like that? Where else has it ever happened? Mum confessed,

'You know, I hate flowers and trees now.' She came out with that and shocked herself, because she'd grown up in a village, and knew them all and loved them. That was before. When we went out for country walks, she could name every flower, every grass and herb: coltsfoot, sweetgrass . . . At the cemetery, they laid a tablecloth on the grass and put out snacks and vodka, but the soldiers checked it with the dosimeter and threw it all away. Buried the lot. The herbs and flowers all set the dosimeters clicking. Where had we taken our grandmother?

I pray for love. But I'm afraid of it, afraid of loving. I have a fiancé, we've handed in our forms at the registry office. Ever heard anything about the Hiroshima *hibakusha*? The people who survived Hiroshima? They can only count on marrying each other. It doesn't get written about or discussed here, but we exist. The Chernobyl *hibakusha*. He took me home, introduced me to his mother. She is a good mum. She works at a factory as a financial manager. A community activist. Goes on all the anti-Communist demos, reads Solzhenitsyn. And this good mother, when she found out I was from a Chernobyl family, one of the evacuees, she asked in surprise, 'But surely you can't have children, my dear?' We'd already handed in our forms at the registry office. He pleaded with me, 'I'll move out of home, we can rent a flat.' But the words rang in my ears: 'My dear, for some people procreation would be a sin.' The sin of loving.

Before him, I had another boyfriend. An artist. We wanted to get married too. It was all going well, until one incident. I went round to his studio and heard him on the phone, shouting, 'What a stroke of luck! You've no idea how lucky you are!' Normally he was so calm, stolid, even; he didn't speak with exclamation marks. Then all of a sudden! What had happened? His friend lived in a student hostel. He'd looked into the room next to his and found a girl hanging. She was strung up by a stocking from the top window. His friend took her down and phoned for an ambulance. But my boyfriend was trembling and spluttering. 'You just can't imagine what he saw, what he's experienced! He carried her in his arms. She had white foam on her lips.' He didn't speak about the girl who had died, didn't feel sorry for her. All he wanted was to see her and memorize it, and then to draw the image. It all came flooding back: how he'd asked me about the colour of the fire at the atomic power station, whether I'd seen cats and dogs being shot, how they were lying in the streets. Had people cried? Did I watch them die?

After that incident, I couldn't stay with him any longer, couldn't answer his questions. (*After a moment of silence.*) I don't know whether I want to meet you again. It feels as though you are looking at me the way he did. Just observing me, memorizing me. Some kind of experiment is being done on us. Everybody is taking an interest. I can't shake off that feeling. Have you any idea why we've had this sin visited on us? The sin of procreation? It's not as though I'm guilty of anything.

Is it my fault if I want to be happy?

Katya P.

Monologue on what St Francis preached to the birds

This is my secret. No one else knows about it. I've only spoken about it to my friend.

I am a cameraman. I went out there, my head filled with what they'd taught us: you only become a real author in war, and all that. My favourite writer was Hemingway, my favourite book *A Farewell to Arms*. I arrived, and people were digging in their plots, there were tractors and seed drills out in the fields. It wasn't obvious what to film. Nothing was blowing up anywhere.

The first shoot was in the village hall. They'd put a television on the stage and brought in the local people. They all listened to Gorbachev saying everything was fine, it was all under control. The village where we were filming was undergoing decontamination. They were washing the roofs, trucking in clean soil. But how do you wash an old woman's roof if it's leaking into her house? The soil had to be stripped off to a spade's depth, meaning the whole fertile layer. Underneath, we just have yellow sand. So this old woman was following the village soviet's instructions, shovelling away the soil, but raking off the manure to keep. Pity I didn't film that. Wherever you went, you'd hear, 'Hey, the movie makers are here. We'll find you some heroes to film.' Heroes like the old man and his grandson who spent two days herding the collective-farm cows from Chernobyl itself. When we'd finished filming, the livestock manager took me to this giant trench where they were burying those same cows with a bulldozer. It never occurred to me to film it. I stood with my back to the trench and shot an episode in the finest Soviet documentary tradition: bulldozer drivers reading their copy of *Pravda*, with the banner headline: 'The Country Stands with You'. Oh, and I got lucky: I looked and there was this stork landing on a field. A symbol! No matter what disaster struck, we would triumph! Life would go on . . .

The country lanes. They're full of dust. By now, I've realized that it isn't just dust: it's radioactive dust. I put the camera away to protect the lens. It was a horribly dry May. How much of that dust we swallowed, I really can't say. A week later, my lymph nodes swelled up. But we scrimped on film, which was as precious as bullets, because First Secretary of the Central Committee Slyunkov was due to arrive. Nobody announced in advance where he was turning up, but we could guess. One day, for example, we drove down the road through clouds of dust, then the next day they were busy laying tarmac, and a good two or three layers thick! So it was easy to tell that was where the Party bigwigs were expected! Later, I filmed them walking terribly carefully across the fresh tarmac. Slap bang down the middle! I caught that on film too, but didn't use it.

No one realized what was going on, that was the most terrifying thing. The radiation technicians were providing one set of figures, while the papers were printing another. Aha . . . that's when the penny began to drop. Nooo! I had a small child, a wife I loved . . . What an idiot I'd been to come here! So they'd give me a medal, but my wife would leave me . . . It was humour that saved us. We went overboard with the jokes. A tramp moves into an abandoned village, where just four old women are left. Someone asks them, 'How is your old fellow?' 'Ah, what a stud! He even goes running off to the other village.' You can't be too deep about it all. This is where we all are. Yes, you've got it: Chernobyl. Though there's a road sweeping ahead, a stream flowing, running its course . . . But it happened! Butterflies fluttering about, a pretty woman standing at the riverside . . . But it happened! I felt a bit like that when someone close to me died. There's the sun. You can hear music drifting over a wall, swallows flitting about under the eaves . . . But he's dead! The rain's falling . . . But he's dead! Do you see? I want to capture my feelings in words, convey what it felt like inside me at the time. Slip into another dimension . . .

I saw an apple tree in blossom and started filming it: the bumble-bees buzzing, the bridal white colour. And there were people working, the orchards were blossoming. I held the camera, but couldn't understand. Something was wrong! The exposure was correct, the picture was beautiful, but there was something not right. Then suddenly it hit me: I couldn't smell a thing. The orchard was in blossom, but there was no smell. It was only later I learned that the body reacts to high radiation levels by blocking certain senses. My mum was seventy-four and, now I thought about it, she complained of losing her sense of smell. So then, I thought, now it's happening to me. I asked the others in my group, there were three of us, 'Does the apple blossom smell?' 'You're right, it doesn't smell of anything.' Something had happened to us. The lilac didn't smell either. Even the lilac! And this sensation came over me that every-thing around was fake. I was in the middle of a stage set. And my mind wasn't in a fit state to get to grips with this, it had nothing to fall back on. There was no map.

Something from my childhood. My neighbour, a former parti-san, described how during the war their unit broke out of an encirclement. She was carrying her little baby in her arms, just one month old, as they went through a swamp with enemy forces all around. The baby was crying. He could give them away, then they'd all be found, the whole unit. So she smothered him. She spoke about it with this odd detachment, as though it had been some other person that did it, and the baby hadn't been hers. I can't remember any more why she brought it up. But I remember very clearly something else, my own horror. What had she done? How could she? I'd thought that the entire partisan unit was breaking out of the encirclement for the sake of the baby, to save it. But here they were smothering an infant to keep those strong and healthy men alive. So what is the meaning of life, then? I didn't want to carry on living after that. It made me – I was just a boy at the time – feel awkward looking at that woman, because of the shocking truth I'd learned about her. And I'd found out something terrible about human beings in general. So how did she feel whenever she looked at me? (*He falls silent for a while.*) See, that's why I don't want to remember it. Those days in the Zone. I'll think up all these different excuses for myself. I don't want to open that door. While I was there, I wanted to understand where I was real and where I was a sham. I'd already had children. My first was a son. When my son was born, I stopped fearing death. The meaning of my life was revealed.

At night, I woke up in the hotel: there was a humming outside the window, these mysterious blue flashes. I pulled back the curtains: dozens of jeeps with red crosses and lights flashing were moving down the street. In total silence. I experienced something like trauma. These scenes from a film popped up in my memory. I was taken straight back to my childhood. We post-war children loved watching war films. We loved scenes just like this, that childish feeling of dread. All of your side has fled town, and you are left alone, with some decision you have to make. What's the right thing to do? Play dead? Or what? And if there's something you have to do, then what is it?

In the centre of Khoyniki, there was a board of honour displaying the top workers. The finest people in the district. But it was a drunkard truck driver who went into the contaminated zone and rescued the children from the kindergarten, not some guy with his photo on the board of honour. Everyone showed their true colours. And then there was the evacuation. First they brought out the children. They loaded them into big Ikarus buses. I caught myself filming things exactly how I had seen them in the war films. And just then, I noticed I wasn't alone: the other people involved in all this activity were behaving the same way. They were acting as if they were in everyone's favourite movie, you know the one, *The Cranes Are Flying*: the odd tear in the eye, a few words of farewell. A wave of the hand. It turned out we were all searching for some form of behaviour that we were already familiar with. We were trying to conform to something. A girl was waving to her mum as if to tell her everything was all right, she'd be brave. We would be victorious! We . . . Because that's what we were like.

I thought: what if I went back to Minsk and everyone was being evacuated there too? How would I say goodbye to my loved ones – my wife and son? I imagined myself, among other things, offering this gesture: 'We'll be victorious! We are fighters.' From as far back as I remember, my father wore army clothes, though he wasn't in the military. Thinking about money was petty and bourgeois, and thinking about your own life was unpatriotic. Hunger was the normal state of things. Our parents' generation had survived the ravages of war, and that's what we needed to do. Otherwise, you'd never become a real man. We were taught to fight and survive in any conditions. As for me, after doing my compulsory service in the army, civilian life seemed bland. At night, a whole band of us used to roam the streets in search of kicks. As a boy, I read a marvellous book called *The Cleaners* – I've forgotten who wrote it – where they went off catching saboteurs and spies. Thrills and chases! That's how we're wired. If you plod off to work and eat well every day, you'll get bored and restless!

We were living in the hostel for some vocational college along with the clean-up workers. They were young guys. We were issued a suitcase of vodka, to draw out the radiation. Suddenly, it turned out we had a medical service team in the same hostel. Made up entirely of girls. 'Right, let's have some fun!' the lads said. Two of them headed off and came straight back with their eyes on stalks! They called us over to take a look. What an eyeful: the girls were walking down the corridor. To go with their tunics, they'd been issued trousers and long johns with these cords dragging along the floor, hanging all loose, and none of them showed the slightest embarrassment. All of it was old, well-worn, and far too baggy. Hanging on them like sacks. Some were in slippers, some in misshapen boots. And over their tunics, they had on these rubber boiler suits impregnated with some chemical coating. Ugh, the smell! Some of them even slept all night in that get-up. Dreadful! And they weren't nurses at all, they were students taken from the college's military training department. They'd been promised it was for two days, but when we arrived, they'd already been there for a month. They told us that they were taken to the reactor, where they'd seen a lot of people with burns, but I only heard about those burns from them. I can still see them in front of me: drifting around the hostel in a trance.

The newspapers wrote that we were lucky the wind hadn't blown the wrong way. Meaning towards the city, towards Kiev. No one knew at the time, no one guessed that it had blown over Belarus. Over me and my little son, Yura. That day I'd gone for a walk with him in the forest. We were picking wood sorrel. My God, why did nobody warn me?

After the trip, I returned to Minsk. I was on a trolleybus, travelling in to work, when I heard snatches of a conversation: some people had been filming in Chernobyl, and their cameraman had died out there. He'd been fried. So I began wondering, 'Who could that be?' I carried on listening: he was a young guy with two kids. They mentioned his name: Viktor Gurevich. There was a cameraman with that name, a very young guy. So he had two children, did he? Why on earth had he never told us? We were approaching the film studio, when somebody corrected them: no, it hadn't been Gurevich, it was Gurin, Sergey Gurin. My God, that was me! It seems funny now, but at the time, I walked from the metro station to the film studio, and was frightened of opening the door and finding . . . I had the ridiculous thought: 'So where did they get a photo of me? The personnel department?' Where had this rumour sprung from? There was a discrepancy between the scale of what was happening and the death toll. If you take the Battle of Kursk, there were thousands of victims. That makes sense. But here, in the first days of the disaster, it was supposedly just seven firemen. Later, a few more people. And then the terms became too abstract for our minds: 'in a few generations', 'eternity', 'nothing'. Rumours began of three-headed birds flying around, hens that were pecking foxes to death, bald hedgehogs.

What came next? They needed to send someone to the Zone again. One cameraman brought in a certificate saying he had a stomach ulcer, another cleared off on holiday. They called me in: 'You have to do it!' 'But I've just got back.' 'Yes, but you've already been out there. So it won't really matter. And, see, you've already got children. The other guys are still young.' Damn it, maybe I'd like a few more, maybe I want five or six kids! They began piling on the pressure, told me there was a pay review coming up and I'd have a trump card to play. They'd raise my salary. It's a sad, ridiculous story. I pushed it into some dark recess of my mind.

Once I filmed some concentration-camp victims. Those people generally try to avoid each other. There's something unnatural about getting together and remembering the war. Dragging up memories of how they killed and were killed, people who had known humiliation or experienced it together. Those people run away from each other. They're running away from themselves. They're running from what they've discovered about human beings: from the things that surfaced there, that burst out of their skin. Right. And here's why. Something was going on there in Chernobyl. I too discovered things, felt things that I don't want to talk about. For instance, the fact that all our humanist views are relative. Under extreme conditions, a person essentially shows himself to be nothing like the characters they write about in books. I've never actually found a person like those literary characters, never run into one. It's the opposite. Man is no hero. We are all peddling the apocalypse. Some in a bigger way, some in a smaller way. There are memories and images that flash through my mind. The chairman of the collective farm wanted two trucks to evacuate his family and all their things, their furniture. The Party organizer asked for one for himself, demanding fair play. Meanwhile, for many days – and I was a witness to this – they were unable to evacuate small children, a nursery group, because of the shortage of transport. And here's this man who can't cram all his household junk into two trucks: his three-litre jars of jams and pickles, even. I saw them loading it up the next day. I didn't film that either. (*Suddenly, he starts laughing.*) We bought sausage and tins of food in a shop out there and were too scared to eat it. We carried it with us in our string shoppers, couldn't bring ourselves to throw it away. (*Becoming serious.*) The wheels of evil will keep turning even during the apocalypse. That's what I've realized. People will keep gossiping, kowtowing to their bosses, rescuing their TV sets and astrakhan coats. Just before the world ends, human beings will be exactly the same as they are now. The same as always.

I feel awkward that I never managed to secure any benefits for my film crew. One of our guys needed an apartment, so I went to the trade union committee. 'Help us out, we spent six months in the Zone. We're supposed to be entitled to benefits.' 'Okay,' they told me. 'Bring your certificates. We need stamped certificates from you.' But while we were out there, we drove to the District Committee, and all we found was a woman called Nastya mopping the corridors. Everyone else had fled. We have one director here who has a whole stack of certificates proving where he was and what he filmed. What a hero!

In my mind I have a great long film that never got shot. It's a whole series. (*He is silent.*) We are all peddling the apocalypse.

We entered a cottage with some soldiers. An old woman lived there alone.

'Well, let's get moving, Granny.'

'Let's go, then, lads.'

'You get yourself ready now, Granny.'

We waited in the street, had a smoke. And then the old woman came out holding an icon, a cat and a tied-up bundle. That was all she was taking.

'You can't take the cat, Granny. It's not allowed. His fur is radioactive.'

'No, lads, I won't go without the cat. How can I leave him behind? He'd be all alone. He's family to me.'

That's where it all began, with that old woman, and the apple tree in blossom. These days, I only film animals. Like I said, the meaning of my life was revealed.

Once I showed my Chernobyl work to some children. I got told off for that, asked why I was doing it. They said that you couldn't do that, it was better not to. So you had these kids living in a state of fear, with all these conversations going on over their heads, with their abnormal blood tests, their immune systems all messed up. I was hoping I might get an audience of five or ten. It was a full house. They asked all sorts of questions, but one in particular remains engraved in my memory. This boy, who was stammering and blushing, obviously a shy character, not a talkative type, asked: 'Why couldn't you help the animals that were left behind?' Well, why indeed? It was a question that simply hadn't occurred to me. And I couldn't answer him. Our art is only about human suffering and love, not about everything living. It's only about the human dimension! We don't want to lower ourselves to the level of the animals, the plants, to that other world. But at the same time, man has the capacity to destroy just about everything. We can kill every living thing. It isn't a fantasy any more. I was told that, in the initial months after the accident, when they were discussing the idea of resettling people, they considered a project for evacuating the animals too. But how? How could you move them all out? You might be able to herd out all the animals on the land, but what about those inside the earth: the beetles and worms? And the ones up above us, in the sky? How do you evacuate a sparrow or a pigeon? What do you do with them? We wouldn't have the means to convey the information they'd need.

I want to make a film. It'll be called *Hostages*. It'll be about the animals. Remember that song, 'A russet island sailing in the sea'. When the ship was sinking, people climbed into the lifeboats. But the horses didn't know that there was no space in the boats for them.

My film's a modern parable. The action takes place on a distant planet. An astronaut is in his spacesuit. Some noise is coming through his headphones. He sees something vast moving towards him. It's immense. A dinosaur? Without understanding what it is, he shoots it. A moment later, again there's something coming towards him. He destroys that too. Then, an instant later, there's a whole herd. He causes a bloodbath. But he discovers a fire had broken out, and the animals were just fleeing along the path where the astronaut was standing. Human beings!

And what happened to me . . . I'll tell you about it. Something unusual. I began seeing animals with new eyes, the trees, the birds. I've been travelling into the Zone all these years. From some abandoned, derelict human house this wild boar will leap out, or an elk. That's what I have filmed. That's what I'm looking for. I want to make a new film, looking at everything through the eyes of the animals. People ask me, 'Why are you filming that? Look about you, there's a war raging in Chechnya.' But St Francis used to preach to the birds. He spoke to them as equals. What if it was the birds that were speaking to him in their bird language, rather than him condescending to their level. What if he knew their secret language?

Remember, in Dostoevsky, a man lashing a horse 'on its meek eyes'. A madman! Not on its rump, 'on its meek eyes'.

Sergey Gurin, cameraman

Monologue without a title: a scream

Good people, leave us be! Back off! You want to have a talk and then go; but we're the ones that have to live here.

I've got these medical records. Every day, I pick them up and read them.

Anya Budai. Year of birth: 1985. 380 rem.
Vitya Grinkevich. Year of birth: 1986. 785 rem.
Nastya Shablovskaya. Year of birth: 1986. 570 rem.
Alyosha Plenin. Year of birth: 1985. 570 rem.
Andrey Kotchenko. Year of birth: 1987. 450 rem.

A mum brought in one such girl today for her appointment.
'What's the trouble?'
'I'm aching all over, just like my granny: my heart, my back, I feel dizzy.'

Since they were little, they've known the word 'alopecia', because a lot of them are bald. They don't have hair: they've got no eyebrows or eyelashes. They're all used to it. But in our village, there's only a junior school. Once they reach fifth grade they get bussed ten kilometres away. They cry, they don't want to go. The other children will laugh at them there.

You've seen it for yourself. I've got a corridor full of patients. They're all waiting. Every day, I hear things that make a nonsense of all your scary films on television. So you can pass that on to the authorities in the capital. A nonsense!

Modern, postmodern . . . In the night, I was called out on an urgent case. I got there . . . The mother was kneeling by the bed: her child was dying. I could hear her wailing: 'If it was going to happen, my little one, I wanted it to be in the summer. It's warm in summer, there are flowers, the ground is all soft. But it's winter now. Try and hold on at least till the spring.' Can you write that sort of thing?

I don't want to trade in their misery, to philosophize. For that, I'd need to stand back, and that's something I just can't do. Every day, I get to hear what they say. Hear them complaining and crying. Good people, do you want to know the truth? Then sit here by my side, writing it all down. But nobody would read a book like that . . .

Better just to leave us be. We're the ones who have to live here.

Arkady Pavlovich Bogdankevich, rural medical assistant

Monologue in two voices: male and female

> *Teachers Nina Konstantinovna Zharkova and Nikolai Prokhorovich Zharkov, husband and wife. Nikolai teaches design and technology and Nina teaches language and literature.*

Her voice:
I spend so much time thinking about death that I don't like going and looking at it. Have you ever heard children talking about death?

You get it in my class. Fourteen-year-olds are already debating and discussing whether dying is scary or not. Not so long ago, little children would have been wondering where they came from, where babies came from. Now they're worried about what might happen after a nuclear war. They've lost their love of the classics. I recite Pushkin to them: their eyes are cold and detached. There's an emptiness . . . It's already a different world around them. They read science fiction, it captivates them. There you'll get a man hopping off the earth, operating in cosmic time, in other worlds. They aren't able to fear death the way adults fear it, people like me, for instance; they're excited by it as something fantastical. A shift to a new place . . .

It's something I ponder. Having death around makes you do a lot of thinking. I teach Russian literature to children who aren't like children used to be ten years ago. Right before their eyes, there's always someone or something being buried. Being laid to rest in the ground . . . People they knew. Houses and trees, everything is being buried. In assembly, the children faint if they remain standing for fifteen or twenty minutes, their noses start bleeding. There's nothing that can surprise them, and nothing that can cheer them up. They're always sleepy and tired, their faces are all pale and grey. They don't play or muck around. And if a fight breaks out, or they smash a window by accident, the teachers are actually rather delighted. They won't tell those kids off, because they aren't like proper children. And they develop so slowly. If you ask one of them to repeat something in class, the child won't be able to do it – to the point where you'll say a sentence, and ask them to repeat it straight after you, and they just can't do it. 'Hey, is anybody home? Hello?' you say, giving them a little shake. I think about things. Do a lot of thinking. It feels like painting on glass with water, I'm the only one who knows what I'm drawing, nobody else sees it, nobody can guess what it is. Nobody can imagine it . . .

Our life revolves around one thing: Chernobyl. Where were you at the time, how far from the reactor did you live? What did you see? Who died? Who's left? Where did they move to? In the first months, I remember the restaurants coming back to life, parties starting up again: 'You only live once', 'May as well go out with a bang'. Soldiers and officers poured into the place. We'll never be rid of Chernobyl now. A pregnant young woman all of a sudden died. No diagnosis; even the pathologist couldn't establish the cause. One little girl hanged herself. She was twelve. Completely out of the blue. Her parents are going crazy. There's a single diagnosis for everything: Chernobyl. No matter what it is that's happened, everyone will say: 'Chernobyl'. People criticize us. 'Your illnesses are all because you're frightened. They come from fear. It's radio-phobia.' Then why are little children getting sick and dying? They don't know what fear is, they're too young to understand it.

I remember those early days. The burning throat, the heaviness, my whole body felt heavy. 'You're imagining things,' the doctor said. 'Everyone has started imagining things because of Cherno-byl.' This isn't imaginary! My body is aching all over, I have no strength. My husband and I feel uncomfortable about admitting it to each other, but our legs have started going numb. Everyone else was complaining about it too, all sorts of people. You're walking along the road and suddenly feel like you could just lie down right there. Lie down and go to sleep. Our pupils would put their heads down on the desks, they were falling asleep during lessons. And everyone became terribly cheerless and gloomy; you could go the whole day without seeing a single kindly face, nobody smiling. From eight in the morning until nine at night, the children were in school. Playing outdoors and running around were strictly forbid-den. They were issued clothing: the girls were given skirts and blouses, the boys had jackets and trousers, but they went home in those clothes and who knows where else. According to the instruc-tions, their mums should have been washing the clothes at home each day, so that the children would come to school in clean things. First, they only gave them one of each item: one blouse, say, and one skirt, and they didn't give them a change of clothing. And second, their mums were overworked with domestic chores – their hens, a cow, a pig – and they didn't understand, in any case, that those things needed to be washed each day. Dirt to them meant ink stains, soil or grease marks, not the effects of some short-lived iso-topes. When I tried to explain certain things to the parents of our pupils, my impression was that they could understand me about as well as if some witch doctor from an African tribe had suddenly turned up. 'So what is it, this radiation stuff? I can't hear it, can't see it. I'm struggling to make ends meet from one payday to the next. By the last three days, we're always down to milk and potatoes.' The mum will pooh-pooh it. But milk isn't allowed, potatoes aren't either. They delivered Chinese tins of stew and buckwheat to the shop, but what can you buy them with? They gave out compensation; it's to compensate us for living here, but it's peanuts. Enough for just two tins. Their safety instructions were designed for someone educated, for a certain lifestyle. But we don't have that here! We don't have the sort of people those instructions were written for. What's more, it isn't all that easy to explain to everyone the difference between rems and roentgens. Or the theory of low doses.

From my own point of view . . . What I'd bring up is our fatalism, a certain tendency towards fatalism. For instance, in the first year, you couldn't use anything from the vegetable plots, but they ate it anyway, bottled it up for the winter. And it was such a marvellous harvest! You try telling people that they can't eat the cucumbers and tomatoes! Can't eat them? What do you mean? They taste normal enough. And they eat them, and don't get a tummy ache. Nobody starts glowing in the dark. Our neighbours laid a new floor that year from the local forest, they tested it: the background radiation level was a hundred times over the safety limit. Nobody tore up those floorboards, they just carried on living there. Everything will sort itself out, they reckon. It will all come good, all by itself, without any help from them, without their need to get involved. At first, some foodstuffs were taken to the radiation monitoring technicians and tested: they were dozens of times higher than the safe levels, but then they stopped doing it. 'I can't hear it, can't see it. Oh, these scientists, they're making it all up!' It all carried on as usual: they ploughed and planted and picked things. The unthinkable happened, and people went on living as they had before. And scrapping the cucumbers from your own vegetable plot was a bigger deal than Chernobyl. Children were kept in school all summer long; the soldiers washed the building with laundry powder, they removed a layer of the surrounding soil. But what about autumn? When autumn came, they sent the pupils to pick the beetroots. And they brought students to the fields, kids from the vocational college. They brought in everyone. Chernobyl wasn't as terrible as the thought of leaving undug potatoes in the fields.

Who is to blame? Well, who can we blame other than ourselves?

Before, we never noticed the world around us. It was like the sky, like the air. As if someone had handed it to us forever and it wasn't dependent on us, it would always be there. Back then, I used to love lying down on the grass in the forest and admiring the sky. It felt so good that I could forget my own name. But now? The forest is beautiful, teeming with bilberries, but nobody ever picks them. In the autumn forest, you rarely hear a human voice. Our sensations are tinged with fear, on a subconscious level. All we've got left is the television, our books. Our imaginations . . . Children are growing up inside their homes. Without the forest and rivers. Looking out of their windows. And they are completely different children. I come to class: 'A doleful time of year the eye yet finds enchanting.' It's always Pushkin, who to my mind is eternal. Sometimes I have a blasphemous thought: what if our entire culture is nothing but a chest full of old manuscripts? Everything that I love . . .

His voice:

A new enemy has appeared. An enemy that stands before us in a new guise.

But we were brought up in a spirit of war. With a military mindset. We were focused on deflecting and recovering from a nuclear attack. We were going to be countering chemical, biological and nuclear war, not trying to rid the body of radionuclides. Not measuring their build-up, monitoring caesium and strontium levels. It can't be compared with a war, that just isn't accurate, but that is the comparison everyone makes. As a child, I lived through the Siege of Leningrad. It can't be compared. In Leningrad, we were living on the front line, under constant bombardment. And the hunger, the long years of hunger, when people were reduced to their animal instincts, their inner beast. But here you merely have to go outside to find everything growing in the vegetable plot! Nothing has changed in the fields or the forest. It just doesn't compare. But that wasn't what I wanted to say. I've lost my thread. It's gone now . . . Ah, yes. When the shelling started, God forbid! Your death wasn't something in the distant future, it was right here and now, that very minute. In the winter, we had the hunger. We burned the furniture, burned everything wooden in our flat, all our books, I think even some old rags went into the stove. Some man would be going along the street and sit down, and the next day you'd walk past and find him sitting there, he'd frozen to the spot, stayed sitting there for a week, or till the spring came. Till it warmed up. Nobody had the strength to cut him out of the ice. If someone fell in the street, it was rare for people to come over and help them up. Everybody would creep on by. I remember how people weren't walking; they were creeping, moving like snails. It just can't be compared with anything!

My mum was still living with us when the reactor blew up, and she used to say, 'We've already survived the worst, my son. We survived the Siege. Nothing can be more horrific.' That's what she thought.

We were preparing for war, nuclear war, building nuclear shelters. We wanted to hide from the fallout as if it was shrapnel from a shell. But it's everywhere. In the bread, the salt . . . We breathe radiation, we eat radiation. The idea of having no bread or salt, of having to eat anything there was – to the point where you'd boil your leather belt in water, just for the smell, so you could eat the smell – that I could understand. But not this. Everything has been contaminated. Now we're faced with working out how on earth we should live here. In the first months, there was fear, particularly among the doctors and teachers, the intelligentsia; basically, the more educated people. They just dropped everything and left. Although they were intimidated and weren't allowed to go. It was military discipline. Hand over your Party card. But what I want to understand is, who was to blame? To answer the question of how we should live here, we need to know who was to blame. Who was it? The scientists, the staff at the power plant? Or was it us, our whole way of seeing the world? We are unstoppable in our desire for more. Our need to consume. They found the guilty parties: the director, the operators on duty. Science. But answer me this: why are we not battling the automobile as the work of the human mind, yet we're battling the atomic reactor? We're demanding that all atomic power stations should be closed and the nuclear scientists put on trial. We curse them! I worship human knowledge. And everything that's created by man. Knowledge . . . Just knowledge in itself cannot be criminal. The scientists these days are victims of Chernobyl too. I want to live after Chernobyl, not to die after it. I want to know what I can cling to in my faith, what will give me strength.

It's on all our minds here. People's reactions are different now. It happened ten years ago, after all, but they still measure it against the war. The Second World War on the Eastern Front lasted four years. You can think of this as already two of those wars. I can list all the different reactions for you: 'Everything is already over.' 'We'll manage one way or another.' 'It all happened ten years ago: we aren't scared any more.' 'We're all going to die! All of us are going to die!' 'I want to emigrate.' 'They have to help us.' 'Oh, what the hell! We need to get on with life.' I don't think I've left anything out. Those are the things we hear day in, day out, what gets said over and over again. The way I see it, we're just research specimens. An international laboratory in the middle of Europe. There are ten million Belarusians, and more than two million of us are living on contaminated land. Nature's laboratory. Come, record the data, do your experiments. They come here from everywhere, from all over the world. They write theses and monographs on us. From Moscow and St Petersburg; from Japan, Germany, Austria. They come here because they're afraid of the future . . . (*There is a long pause in the conversation.*)

What was I thinking? I was drawing comparisons again. I was thinking about how I can talk about Chernobyl, but not about the Siege. I got a letter from Leningrad. Sorry, but the name 'St Petersburg' just won't stick in my mind, because Leningrad was the city where death came close. Right, then . . . This letter was an invitation to a gathering of children of the Siege of Leningrad. I went along to it, but I couldn't get out a word about myself. Should I have just talked about fear? It wouldn't have been enough. Just about fear . . . What did it do to me, that fear? I still don't know . . . At home, we never really talked about the time of the Siege. My mother didn't want us to bring it up. But we do talk about Chernobyl. No. (*He stops speaking.*) We don't talk about it among ourselves, but the topic comes up when someone visits us: foreigners, journalists, relatives, people who don't live here. Why don't we talk about Chernobyl among ourselves? It isn't an issue for us. Not with the pupils at school, not at home. We block it out, the subject's closed. People talk about it with the kids while they're in Austria, France, Germany for their treatment. I ask the children what the people there want to know about, what they're interested in. But often they can't even remember the city or village, or the surnames of the people who hosted them. They'll list the presents, the nice food they ate; someone got a tape recorder, someone didn't. They'll come back in new clothes they didn't work for and their parents didn't either. It's a bit like they've been at an exhibition, visited some big department store, a luxury supermarket. They can't wait to be taken back, to be shown more things and given more gifts. They get used to it. They're already used to it. This is the way they live, their vision of life. After their big shopping expedition that goes by the name of 'abroad', after their lavish exhibition, I have to go and face them in the classroom for lessons. I go in, and I can see they've become spectators. They just look, instead of living. I've got to help them. I've got to explain to them that the world isn't one big supermarket. It's something else, much harder work and more beautiful. I take them to my studio, where I have my wooden sculptures. They like them. I tell them, 'You can create all of that from an ordinary piece of wood. Have a go for yourselves.' Come on, wake up! It helped me to get myself out of the Siege; it took me years.

The world has split in two: there is us, the people of Chernobyl, and you, everyone else. Have you noticed? We don't make a point of this 'I'm Belarusian', 'I'm Ukrainian', 'I'm Russian'. Everyone just calls themselves Chernobyl people. 'We're from Chernobyl.' 'I'm a

Chernobyl person.' As if we're some sort of separate people. A new nation . . .

Monologue on how some completely unknown thing can
worm its way into you

Ants . . . Tiny ants crawling along a tree trunk.

You have these army vehicles rumbling all around. Soldiers. Shouting, squabbling, cursing. Helicopters whirring. And there they are, crawling along. I was returning from the Zone and, of all the things I saw that day, that image was the one clear memory I was left with. That moment . . . We had stopped in the forest, I got up to have a smoke near a birch tree. I stood there, leaning against it. Right up close to my face were those ants crawling on the trunk, deaf to our presence, not paying us the slightest attention. Stubbornly following their path. We could vanish and they wouldn't notice. Something along those lines flashed through my mind, among my random thoughts. I was so overloaded with impressions that I couldn't think. I was watching them. I had never really noticed them before. Not close up . . .

At first, everyone called it a 'disaster', then 'nuclear war'. I'd read about Hiroshima and Nagasaki, seen the documentary footage. It was horrific, but everything made sense: a nuclear war, a blast radius. I could imagine all that. But what had happened to us . . . It was simply beyond me. Beyond my knowledge, beyond all the books I'd read in a lifetime. I came back from the trip, and in my bewilderment I scoured the bookshelves in my study. I began reading. But I might as well have not bothered. Some completely unfathomable thing was destroying the whole of my previous world. Here it was, creeping and worming its way into you, and there was nothing you could do . . . I remember having a conversation with a scientist. 'This will be for thousands of years to come,' he was telling me. 'Uranium decays at 238 half lives. Translated into time that means one billion years. And thorium is fourteen billion years.' Fifty years, 100 . . . 200 . . . But after that? After that it's a blank wall, a shock! I lost all sense of what time means, of where I was.

To write about it when just ten years have passed . . . a mere blink of an eye . . . Writing now? I think it's risky! Too unreliable. In any case, we'll just be inventing some vague semblance of our lives. Tracing out a copy. I tried, and it didn't work. After Chernobyl we were left with the whole mythology of Chernobyl. Newspapers and magazines compete to write the most terrifying stories. People who weren't there particularly love the hair-raising stuff. We've all read about the mushrooms the size of a human head, but nobody has ever found one. Or the birds with two beaks. That's why what we need isn't to write, but to record. To document it all. Give me a sci-fi novel about Chernobyl. They don't exist! And no one will write one, you can be sure of that. No one.

I have a special notebook, I started keeping notes in it from the early days. I recorded conversations, rumours, jokes in it. That's the most interesting and authentic thing of all. An accurate impression. What do we have left of the ancient Greeks? The myths.

I'll give you that notebook. It'll just end up lying among my papers. Well, maybe I'll show it to my children when they grow up. It is history, after all.

From conversations

For a third month they are telling us on the radio, 'The situation is stabilizing . . . The situation is stabilizing . . . The situation is stabilizing . . .'

They instantly revived the forgotten vocabulary of Stalinism: 'Western intelligence agents', 'arch enemies of Socialism', 'spying forays', 'sabotage', 'a stab in the back', 'subverting the inviolable union of Soviet peoples'. Everybody is harping on about undercover spies and saboteurs, rather than iodine prophylaxis. Any unofficial information is treated as enemy ideology.

In my news report yesterday, the editor cut the story about the mother of one of the firemen who put out the nuclear fire on the first night. He died of acute radiation sickness. After burying their son in Moscow, the parents came back to their village, which had already been evacuated.

They secretly made their way home through the forest and picked a sackful of tomatoes and cucumbers. The mother was happy: 'We bottled up twenty jars' worth.' Their trust in the land, in the age-old peasant experience . . . Even the death of their son could not shake their familiar world.

My editor called me in. 'Been listening to Radio Liberty?' I said nothing. 'I don't need panic-mongers in my newspaper. Go and write about heroes. The soldiers who climbed on the roof of the reactor.'

The hero. Heroes . . . Who are they today? To my mind, they are the doctors who defy orders from above and tell people the truth. And the journalists and scientists. But, as one editor at our editorial meeting said, 'Remember! We have no doctors, teachers, scientists or journalists. We all have one profession now: that of being a Soviet citizen.'

Did he believe his own words? How can he not be scared? My faith is being sapped by the day.

Some political supervisors from the Central Committee turned up. Their itinerary: travel by car from the hotel to the Provincial Party Committee and back to the hotel, again by car. They read up on the situation from files of the local newspapers. Their travel bags are stuffed with sandwiches from Minsk. They brew their tea with bottled water. They brought that along too. I was told that by the receptionist of the hotel where they're staying. People don't believe the newspapers, television or radio: they look for clues in the behaviour of their bosses. That is their most reliable source.

What should you do if you've got a child? I feel like just grabbing mine and running. But I've got a Party card in my pocket. I can't do that!

The fairy tale most popular in the Zone is that Stolichnaya vodka is the best antidote to strontium and caesium.

But suddenly the village shops have filled up with all sorts of goods in short supply. I heard the secretary of the Provincial Party Committee giving a speech: 'We'll create a heaven on earth for you. Just stay here and work. We'll swamp you with sausage and buckwheat. You'll have everything there is to be found in the best restricted-access shops.' (By

which he meant the delicacies sold in their own Committee's canteens.)
The attitude towards ordinary people is: keep them happy with vodka
and sausage.

But damn it, I've never seen three varieties of sausage on sale in a
village shop before! I bought some imported tights there for my wife.

Dosimeters were on sale for a month but then disappeared. We are not
allowed to write about that. The kinds and quantities of radionuclide in
the fallout are also off limits. We aren't allowed to report that only men
now live in the villages: the women and children have been evacuated. All
summer long, the men have been washing their own laundry, milking the
cows, digging in the plots. And, of course, getting drunk and fighting. A
world without women? Pity I'm not a scriptwriter: it would make a great
film. Where is Spielberg when you need him? Or my favourite director,
Alexey German? I wrote about all that, but again the merciless editorial
red pencil struck it out. 'Don't forget we have enemies. Many enemies
across the ocean.' And that's why only good things can happen in the
USSR. Nothing bad, or beyond comprehension.

All the same, special trains have been arriving at certain platforms. Officials have been spotted with their luggage packed.

An old woman stopped me as I was passing a police checkpoint: 'Look in
on my cottage over there, will you? It's time to be digging up the spuds,
but the soldiers won't let me through.' They've been resettled. They were
lied to and told it was only for three days, otherwise they would never
have gone. People who are in limbo, with nothing to their name. They
make their way back to the villages across army barriers, along forest
trails, through the swamps. At night. The authorities hunt them in trucks
and helicopters, try to catch them. 'Like when the Germans were here,'
the old people say. 'During the war.'

I saw my first looter. A young lad dressed in two sheepskin jackets. He
tried to persuade the army patrol it was his way of treating back pain. He
cracked under interrogation. 'It's a bit scary the first time, but you soon
get used to it. Just down a vodka and in you go.' Overriding the survival

instinct. Sober, you could never do it; but this is how we pluck up courage for our daring deeds. And misdeeds.

We went into an empty cottage and found an icon laid on a white table-cloth. 'For God,' someone remarked.

In another, the table had only the white tablecloth. 'For people,' someone said.

Went back to my family village a year later. The dogs had gone wild. Found our Rex and called him, but he didn't come. Didn't recognize me? Or didn't want to? He was sulking.

Those first weeks and months, everybody was very quiet. Nobody spoke. They were in a stupor. They had to leave, but wouldn't face it until the last day. Their brains were switched off. I don't remember anyone talking seriously, only joking. 'Now all the shops have radio-goods'; or, 'Men who can't get it up are either radioactive or radiopassive'. Then suddenly the jokes stopped.

At the hospital, a little girl tells her mother, 'That boy's died, but yesterday he was giving me sweets.'

In the sugar queue: 'Hey, people, I'm telling you, there are so many mush-rooms this year! Mushrooms and berries, like they've been planted specially.'

'But they're contaminated.'

'Don't be daft, you don't have to eat them! Pick them, dry them and take them to market in Minsk. You'll make a fortune.'

Can we be helped? And how? Maybe transplanting our people to Australia or Canada? That's what is reportedly being discussed from time to time at the highest levels.

They used to choose the site for a church by watching the skies. The church people saw signs. Before construction began, they'd perform rites. But an atomic power station got built like a factory, or your average pig farm. The roof was coated in asphalt. Bitumen. It melted when the plant caught fire.

Did you read about it? They caught a runaway soldier outside Chernobyl. He had made himself a dugout and was living near the reactor. He made the rounds of abandoned houses, surviving on pork fat and pickled cucumbers, trapping wild animals. He had deserted when the elder soldiers beat him to a pulp. Chernobyl offered him refuge.

We are fatalists. We don't embark on action, because we believe whatever will be, will be. We believe in destiny; that is our history. Every generation's been plagued with war and bloodshed. How could we be any different? Fatalists.

The first wolf-dogs have appeared, born to she-wolves from dogs that took to the forest. They are larger than wolves, cannot be hunted with flags, are unafraid of light and humans, do not respond to wolf decoys. And the wild cats are already roaming in packs, with no fear of people. Their memory of how they were obedient to man has faded. There is a blurring of the boundaries between the real and unreal.

Yesterday, my father turned eighty. The whole family gathered around the table. I looked at him and thought about how much had been crammed into his life: Stalin's Gulags, the war and now Chernobyl. It all fell in the era of his generation. That one generation. He loves fishing, digging in the garden. As a young man, to Mother's dismay, he was a skirt-chaser: 'There wasn't a woman he didn't run after.' These days, I've noticed he'll lower his eyes whenever a beautiful young woman walks his way.

What do we know about man? About what he is capable of? . . . About what he has in him? . . .

From the grapevine

They're building camps outside Chernobyl to hold all the people who were exposed to radiation. They'll hold them there, watch them and bury them.

They are already bussing the dead from villages near the power station straight to the cemetery, burying thousands in mass graves. Like in the Siege of Leningrad.

Some people supposedly saw a mysterious glow in the sky above the power station the day before the explosion. Somebody even took a photo. The negative revealed a sort of hovering extraterrestrial object.

In Minsk, they've washed down the passenger and goods trains. They're going to move the entire capital out to Siberia. The old barracks from Stalin's Gulags are being fixed up. They'll start with the women and children. The Ukrainians are on their way there already.

Anglers are finding more and more amphibious fish that can live in water and on land. On the land, they walk on their fin-paws. They've begun catching headless and finless pike: just a swimming belly.

Something of the kind will start happening to people soon. The Belarusians will morph into humanoids.

It wasn't an accident: it was an earthquake. Something took place in the subterranean crust. A geological explosion. Geophysical and cosmophysical forces were at work. The military knew in advance and could have warned us, but it was all strictly classified.

The forest animals have radiation sickness. They wander around dejectedly, with sadness in their eyes. The hunters feel too scared and sorry to shoot them. And the animals have lost their fear of humans. Foxes and wolves come into the village and nuzzle up to children.

Chernobyl people can bear children, but they'll have an unknown yellow fluid pulsing through them instead of blood. There are scientists who argue that the monkeys became so clever because they lived in radiation hotspots. Children born in three or four generations will all be Einsteins. A cosmic experiment is being conducted on us.

Anatoly Shimansky, journalist

*Monologue on Cartesian philosophy and on eating a
radioactive sandwich with someone so as not to be ashamed*

I lived among books. For twenty years, I lectured at a university . . .

I am an academic. A man who picked out his favourite period in history and resides there. Totally preoccupied with it, immersed in his own space. In a perfect world . . . That was how it would have been ideally, of course. Because, at that time, the philosophy we had was Marxist-Leninist, and the topics on offer for a thesis were the role of Marxism-Leninism in the development of agriculture or in clearing virgin lands. Or the role of the leader of the world proletariat . . . All in all, they had no time for Cartesian thought. But I was in luck. My undergraduate dissertation was entered into a competition in Moscow, and somebody made a phone call from there, saying: 'Don't touch this fellow. Let him write.' I was working on the French religious philosopher Malebranche, who undertook to interpret the Bible from the perspective of the rational mind. The eighteenth century was the Age of Enlightenment. Faith in reason, the idea that we are capable of explaining the world. As I now realize, I was lucky. I was saved from the mincer. Saved from a lot of aggravation. A miracle! Before that, I was warned more than once that the choice of Malebranche for my dissertation could be seen as interesting, but for my thesis I would have to think carefully about the topic. That was a serious matter. Here they were, they said, allowing me to stay on to do postgraduate work in the department of Marxist-Leninist philosophy, and I was proposing to emigrate to the past . . . Surely I could see the problem . . .

Gorbachev's perestroika began. We had waited so long for this moment. The first thing I noticed was how people's expressions immediately began to change, all of a sudden their faces were different. They even began to walk differently. Life was subtly altering the way they moved. They were smiling at each other more. I picked up on a different energy in everything. Something had changed. Completely. To this day, I am amazed at how quickly it happened, and as for me . . . I was pulled abruptly out of that Cartesian idyll. Instead of books on philosophy, I now read the latest papers and magazines, eagerly awaiting each new perestroika-inspired issue of *Ogonyok*. In the morning, there were queues at the news-stands; never before – or after – had people read the papers with such relish. They would never again believe them so unquestioningly. There was an avalanche of information. Lenin's political testament was published, which had been locked away for half a century in some special archive. The bookshops began stocking Solzhenitsyn, then Shalamov, Bukharin. It wasn't so long ago that you could have been arrested for possessing those books. You could have earned yourself a prison sentence. Andrey Sakharov was brought back from exile. For the first time, they broadcast sessions of the USSR's Supreme Soviet on television. The whole country sat glued to their screens. We talked and talked. You could say out loud things which until recently would only be discussed in the privacy of your kitchen. For so many generations, we had been whispering in our kitchens. So many people went to waste, whiling away their time in dreams, throughout our seventy-odd years of Soviet history. Now, though, everybody was going to rallies and demonstrations. Signing something, voting against something. I remember one historian appearing on television. He brought a map of Stalin's camps to the studio. The whole of Siberia was dotted with red flags. We discovered the truth about Kurapaty . . . What a shock! Society was left reeling! Belarusian Kurapaty was the site of a mass grave in 1937. Belarusians, Russians, Poles and Lithuanians were buried there together, in their tens of thousands. The NKVD's ditches were dug two metres deep, and people were stacked in two or three layers. Once that place had been a long way outside Minsk, but later it fell within the city limits. It became part of the city, you could catch a tram there. In the 1950s, the area was planted with young trees, the pines grew taller, and the city people suspected nothing. They had picnics there at the weekends. In winter, they skied there. People began excavations. The authorities – the Communist authorities – had lied. They tried to wriggle out of it. By night, the police filled the graves back in, but in the daytime people dug them open again. I saw documentary footage: rows of skulls cleaned of soil. And each one had a hole in the back of the head . . .

Of course, we lived with the feeling that we were taking part in a revolution. In a new phase of history.

Don't worry, I haven't digressed from the topic. I want to remember the general mood at the time Chernobyl happened. Because they will always go together in history: the downfall of Socialism and the Chernobyl disaster. They coincided. Chernobyl hastened the collapse of the Soviet Union. It blew the empire apart.

And it made me into a politician.

It was 4 May, day nine after the accident, when Gorbachev made his appearance; it was cowardice, of course. Befuddlement. Like in the early days of the war, in 1941. The newspapers were writing about enemy ploys and Western hysteria, about the anti-Soviet commotion and damaging rumours spread by our overseas opponents. I remember how I felt in those days. The fear didn't set in for a long time: for almost a month everyone was on tenterhooks, waiting for them to announce that, under the leadership of the Communist Party, our scientists, our heroic firemen, our soldiers have once again conquered the elements. They have won an unprecedented victory, they have driven the cosmic fire back into a test tube. The fear took a while to set in. For a long time, we kept it out. Yes, that was it. Absolutely! As I now realize, we could not make the mental connection between fear and peaceful nuclear energy. From all the textbooks and other books we'd read, in our minds we pictured the world as follows: military nuclear power was a sinister mushroom cloud billowing up into the sky, like at Hiroshima and Nagasaki, incinerating people instantly; whereas peaceful nuclear energy was a harmless light bulb. We had a childish image of the world; we were living life as depicted in children's stories. It wasn't just us; the whole of mankind wised up after Chernobyl. We all grew up, became more mature.

A few conversations from the early days:

'There's some nuclear power plant on fire. But it's happening a long way away, in the Ukraine.'

'I read in the papers that we're sending combat vehicles there. The army. We'll overcome it!'

'In Belarus, we don't have a single atomic power station. We aren't worried.'

My first trip into the Zone.

I went there thinking it would all be covered in grey ash, in black soot, like in Bryullov's painting *The Last Day of Pompeii*. But I got there and everything was beautiful. Breathtakingly beautiful! Meadows in flower, the gentle spring green of the forests. I love that time of year. Everything is coming to life. Flourishing, singing . . . What struck me most was the combination of beauty and fear. Fear could no longer be separated from beauty or beauty from fear. Everything was turned on its head, topsy-turvy. I realize that now. There was a strange sensation of death . . .

We arrived as a group. Nobody sent us there. A group of Belarusian deputies from the opposition. What times they were! What times! The Communist authorities were backing down. They were weakening, losing confidence. Everything was fragile, but the local authorities treated us with hostility: 'Do you have permission? What gives you the right to stir up the people? To ask questions? Who gave you this assignment?' They alluded to instructions received from above: 'Do not give in to panic. Await orders.' As if to say, 'Don't you go scaring people when we need to fulfil our quotas, our grain and meat quotas.' What worried them was not people's health, but hitting their targets. The quotas for the republic and the quotas for the nation . . . They were afraid of their bosses. And their bosses were afraid of those above them, and so on up the chain, all the way to the general secretary. One man decided everything from his celestial heights. That was how the pyramid of power was built. It was headed by a tsar. At that time, a Communist tsar. 'Everything here is contaminated,' we told them. 'None of the food you've produced can be eaten.' 'You are rabble rousers. Stop your enemy agitation. We'll make phone calls. We'll report this.' And they made their phone calls. They reported it to 'the appropriate authorities'.

The village of Malinovka. Fifty-nine curies per square metre.

We went into the school. 'So how are you doing?'

'We're all scared, of course. But we've been reassured: all we need to do is wash the roof, close off the wells with plastic, and tarmac the country lanes. Then we can go on living here. Though the cats keep scratching for some reason, and the horses' noses are dribbling mucus on to the ground.'

The head teacher invited me to her home, for lunch. It was a new house. She'd held a housewarming there two months earlier. In Belarusian, it's called 'vkhodiny': when people have only just moved into a house. Near the house, they had a sturdy barn and a root cellar. What was once known as a kulak farmstead. These were the kind of people dispossessed under Stalin's dekulakization. Enough to make you stop, stare and envy.

'But soon you'll have to leave.'

'Out of the question! We've put so much work into this place.'

'Take a look at the dosimeter.'

'They've been coming here, those bloody scientists! They won't let people live in peace!' The husband waved his hand and went off to the meadow to fetch his horse without saying goodbye to me.

The village of Chudyany. One hundred and fifty curies per square metre.

Women were digging in their vegetable plots, children running about the streets. At the end of the village, the men were hewing timber for a new log house. Our car stopped near them. They clustered round, asked for a smoke.

'How are things in the capital? Are you getting vodka? We keep running dry here. A good job we're brewing our own. Gorbachev doesn't touch a drop himself and he won't let us drink either.'

'Aha, so you're deputies. The tobacco situation is pretty lousy around these parts too.'

'Listen, guys,' we began explaining to them. 'You're going to have to leave here soon. See this dosimeter? The radiation where we're standing is a hundred times over the limit.'

'Oh, come on, don't give us that crap. Hell, what do we want with your dosimeter! You won't be hanging around here long, but we've got to live in this place. You can stick that dosimeter of yours where the sun don't shine!'

I've watched the film about the *Titanic* a few times, and it reminded me of what I saw with my own eyes. I experienced it myself, in the early days of Chernobyl. Everything was just like on the *Titanic*. People were behaving in exactly the same way. The psychology was the same. I recognized it. I even made the comparison at the time. You had the bottom of the ship already pierced, this tremendous surge of water was flooding the lower holds, overturning barrels and crates. It's creeping forward, breaking through all obstacles. While up above, the lights are bright, music is playing, champagne is being served. Families carry on squabbling, love affairs are being kindled. And the water is gushing up the staircases, into the cabins . . .

The lights are bright, music is playing; champagne is being served . . .

Our mentality is a separate topic. For us, everything revolves around feeling. That is what gives us our grandness, elevates our lives, and is, at the same time, so disastrous. The rational choice for us is never enough. We gauge our actions with our hearts, not our minds. The moment you wander into someone's yard in the village, you are their guest. They are so pleased. Then they are upset. They will anxiously shake their heads. 'Oh dear, we've no fresh fish, nothing to offer.' 'Perhaps you'd like some milk? I'll pour you a mugful.' They won't just let you go on your way. They'll beckon you into their cottage. Some of the others were afraid, but I was willing. I went in, sat at their table, ate radioactive sandwiches because that was what they were all eating. I downed a drink with them and it gave me a sense of pride to know I had that in me. I had it in me! Yes, that's right! I told myself, 'Okay, so maybe I can't change a thing in this man's life, but what I can do is eat a radioactive sandwich alongside him, so I won't be ashamed. Share his fate.' That is the attitude we take to our own lives. And yet I had a wife and two children. I was responsible for them. I had a dosimeter in my pocket. I realize now, this is just our world, it's who we are. Ten years ago, I felt proud of being the way I was, while today I'm ashamed of it. All the same, I would still sit with him and eat that wretched sandwich again. I've thought about it, thought about what kind of people we are. I couldn't get that damned sandwich out of my mind. You had to eat it as an act of the heart, not the mind.

A writer put it well when he observed that in the twentieth century, and now in the twenty-first, we are living, as we were taught to, by the precepts of nineteenth-century literature. Lord knows, I'm often plagued by doubts. I've discussed this with many people. Who on earth are we?

I had an interesting conversation with the wife, now the widow, of a helicopter pilot. She was an intelligent woman. I sat talking to her for a long time. She wanted to understand too, to understand and find meaning in her husband's death in order to be able to accept it. She just couldn't.

I read many times in the newspapers about the helicopter pilots working above the reactor. At first, they were dropping lead panels into it, but they vanished down the hole without trace. Then someone remembered that lead vaporizes at 700 degrees Celsius and the temperature down there was 2,000 degrees. After that, sacks of dolomite and sand were dropped down. Up where the pilots were was as dark as night from the dust that raised. Pillars of dust. In order to drop their 'bombs' accurately, they opened the cabin windows and aimed by eye to get the correct banking of the helicopter, left–right, up–down. The radiation doses were ridiculous! I remember the headlines of the newspaper stories: 'Heroes of the Sky', 'Falcons of Chernobyl'. And then there was this woman. She admitted her doubts to me. 'Now they write that my husband is a hero. Yes, he is, but what does that mean? I know he was an honest, dutiful officer. Disciplined. He came back from Chernobyl and was ill within a few months. They presented him with an award in the Kremlin, and he saw his comrades there. They were ill too, but glad to have met up again. He came home happy, with his medal. I asked him then, 'Could you not have avoided being so severely affected? Protected your health better?' 'I probably could, if I'd thought more about it,' he said. 'We needed proper protective clothing, special goggles, a face mask. We had none of that. We didn't follow standard safety procedures ourselves. We weren't thinking about that at the time.' Actually, none of us were. What a pity that, in the past, we did so little thinking! From the viewpoint of our culture, thinking about yourself was selfish. It showed a lack of spirit. There was always something more important than you and your life.

1989. The third anniversary of 26 April. Three years had passed since the disaster. Everyone had been evacuated from a thirty-kilometre zone, but over two million Belarusians were still living in contaminated areas. Forgotten. The Belarusian opposition planned a protest for that day, and the authorities responded by declaring a 'volunteer Saturday' to clean up Minsk. They put up red flags, brought out mobile food stalls with delicacies in short supply at the time (fresh smoked sausage, chocolates, jars of instant coffee . . .). Police cars were on the prowl, heavies in plain clothes snooping about taking photos, but – a sign of new times! – people just ignored all that. They were no longer afraid of them. They began assembling at Chelyuskin Heroes Park, more and more of them. By ten o'clock, there were already 20,000–30,000 (I've taken that from police estimates later reported on television), and the crowd was growing by the minute. We hadn't expected that ourselves. Everything was just getting bigger and bigger. Who could stem this tide of people? At precisely ten o'clock, as we'd planned, the march moved off along Lenin Prospect towards the city centre, where the rally was to be held. All along the way, new groups were joining us, waiting for the march on parallel streets, in side streets, in gateways. A rumour spread that the police and army patrols had blocked the roads into the city; they were stopping buses and cars with protesters from other places and turning them back, but no one panicked. People got out and walked on to join up with us. That was announced over a megaphone and a great 'Hurrah!' swept over the march. Balconies were thronged, windows thrown open, people stood on windowsills to wave to us. They were waving shawls and children's flags. Then I noticed, and everybody around started talking about it: the police had melted away – and the boys in plain clothes – taking their cameras with them. I understand now that they were given new orders; they retreated into courtyards, and locked themselves in cars concealed under tarpaulins. The authorities had taken cover and were waiting to see what would happen. They were scared. People marched on in tears, everybody holding hands. They were crying because they were overcoming their fear. They were liberating themselves from the intimidation.

The rally began, and although we'd been preparing for a long time, discussing the list of speakers, it was promptly ignored. People came to the hastily erected platform and spoke without notes, ordinary people from the area around Chernobyl. The list re-formed itself spontaneously. We were hearing witnesses, testimony. The only prominent figure to speak was Academician Velikhov, one of the former directors of the centre in charge of dealing with the accident, but his was not one of the speeches I remember. The ones I do remember are:

A mother with two children, a boy and a girl. She brought them up on to the platform with her. 'My children have not laughed for a long time now. There is no naughtiness. They don't run around in the courtyard. They have no strength. They are like a little old man and woman.'

A woman involved in the clean-up operation. When she pushed back the sleeves of her dress to show the crowd her arms, we saw her sores and scabs. 'I washed clothes for our men working near the reactor,' she said. 'We did most of the laundry by hand because not enough washing machines were delivered, and they soon broke down because they were so overloaded.'

A young doctor. He began by reciting the Hippocratic oath. He talked about how all the data on radiation sickness were being stamped 'secret' or 'top secret'. Medicine and science were being dragged into politics.

This was Chernobyl's public inquiry.

I will not attempt to hide it. I openly admit that this was the most memorable day of my life. We were so happy.

The next day, those of us who had organized the demonstration were summoned by the police and convicted for the fact that a crowd thousands strong had blocked the avenue and obstructed the free movement of public transport. Unauthorized slogans had been displayed. Each of us was given fifteen days under the 'aggravated hooliganism' article of the penal code. The judge passing sentence and the policemen accompanying us to the detention centre were shame-faced. All of them. We were laughing. Yes, because we were so happy!

Now the question was: what more are we capable of? What should we do next?

In one of the Chernobyl-affected villages, a woman fell to her knees in front of us when she heard we were from Minsk. 'Save my child! Take him with you! Our doctors can't understand what's wrong, and he's suffocating, turning blue. He's dying.' (*Falls silent.*)

I went to the hospital. The boy was seven. Thyroid cancer. I wanted to take his mind off it and began joking. He turned to the wall and said, 'Just don't tell me I'm not going to die. I know I am.'

At the Academy of Sciences, I think it was, I was shown an X-ray of someone's lungs that had been burned through by 'hot particles'. They looked like the sky at night. The hot particles were microscopic pieces of radioactive material created when the burning reactor had lead and sand tipped into it. Atoms of lead, sand and graphite combined and were shot high up into the atmosphere. They were dispersed over great distances, hundreds of kilometres. Now they were entering people's bodies via the respiratory tract. The highest mortality was among tractor and truck drivers, people who ploughed the land or drove along the dusty country roads. An organ in which these particles settle glows in X-rays. It is peppered with hundreds of tiny holes, like a fine sieve. The person affected dies, literally burns up; but whereas they are mortal, the hot particles live on. A person dies, and after a thousand years will have turned back into dust. The hot particles, though, are immortal, and their dust will be capable of killing again. (*Falls silent.*)

I came back from those trips . . . I was overwhelmed. I told people what I'd seen. My wife is a specialist in linguistics. She'd never taken an interest in politics before – any more than she had in sport – but now she kept asking me over and over again, 'What can we do? What should we do now?' And we set out on a course which common sense would have told us was impossible. It was the kind of thing a person could only countenance in a time of upheaval, of complete inner emancipation. That was such a time, a time when Gorbachev was making the running, a time of hopes, of faith! We decided to save the children. To reveal to the world the peril Belarusian children were living in. To ask, to shout for help. To raise the alarm. The authorities were silent. They had betrayed the people, but we would not be silent.

Very soon, a group of dedicated helpers and like-minded supporters came together. Our watchword was: 'What are you reading? Solzhenitsyn, Platonov? Welcome!' We were working twelve hours a day. We needed to think of a name for our organization. We went through dozens of possibilities before settling on the simplest: the Children of Chernobyl Foundation. Today, it's impossible to explain or even imagine all our doubts at that time, the arguments, our fears . . . Today, there are countless funds like ours, but ten years ago we were the first: the first civil initiative, unsanctioned by anyone in authority. The response from all officials was identical: 'Foundation? What foundation? We have the Ministry of Health for this sort of thing.'

I understand today: Chernobyl liberated us. We learned to be free.

I remember . . . (*Laughs.*) I can picture it right now! The first refrigerated lorries bringing humanitarian aid drove into the courtyard of our apartment block, to our home address. I looked out my window and saw them, and couldn't imagine how we were going to unload and store it all. The trucks had come from Moldova, with seventeen to twenty tonnes of fruit juice, dried fruit and baby food. By then, there were already rumours that the best way to draw out radiation was to eat lots of fruit, have lots of fibre in your food. I telephoned friends, some at their dachas, others at work. I and my wife began unloading the trucks by ourselves but gradually, one by one, people came out of our block (which was, after all, nine storeys high), and passers-by stopped to ask, 'What are these trucks doing here?' 'They've brought aid for the Chernobyl children.' They dropped whatever they were doing and rolled up their sleeves. By evening, the trucks were unloaded. We packed the goods into cellars and garages, made arrangements with a school. We laughed at ourselves later, but when we brought these gifts to the contaminated areas, when we began distributing them ... people usually assembled in the local school or at the House of Culture. Something's just come back to me now. One time, in Vetka District ... a young family. They, like everyone else, had been given little jars of baby food, cartons of fruit juice. The father sat down and wept. These jars and cartons were too late to save his children's lives. They could make no difference, but he was crying because, after all, they had not been forgotten. Someone had remembered them. There was hope.

The whole world responded. People agreed to take our children for treatment in Italy, France, Germany ... Lufthansa flew them to Germany at the airline's expense. There was competition among the German pilots to come here. We got only the best! As the children were walking out to the aircraft, they all looked so pale, and they were so quiet. There were some odd moments. (*Laughs.*) The father of one boy burst into my office and demanded his son's documentation back: 'They'll take our children's blood! They'll conduct experiments on them!' Of course, the memory of that terrible war still festers. People have not forgotten. But there was something else at work: we had lived behind the barbed wire for such a long time, in the 'Socialist Camp'. We were afraid of that other world. We knew nothing about it.

The Chernobyl mothers and fathers were a different problem. To continue the conversation about our mentality, the Soviet mentality. The Soviet Union had fallen, collapsed, but people were still expecting to be coddled by a great, powerful country, which no longer existed. My characterization, if you want it: a hybrid between a prison and a kindergarten, that's what Socialism is, Soviet Socialism. A citizen surrendered his soul to the state, his conscience, his heart, and in return received his rations for the day. Beyond that, it was a matter of luck: one person got a bigger ration, another a small one. The only constant was that you got it in return for selling your soul. And the thing we most wanted to avoid now was our foundation turning into a distributor of that kind of ration: the Chernobyl ready-packed meal. People were used to waiting and complaining. 'I am a Chernobyl victim. I am entitled, because I am one of the victims.' As I see it today, Chernobyl was a major test of our spirit and our culture.

That first year, we sent 5,000 children abroad. The second year it was 10,000; and in the third, 15,000.

Have you talked to the children about Chernobyl? Not the adults, the children. They often have unexpected ideas. As a philosopher, I'm continually surprised. For example, one girl told me their class was sent out into the countryside in autumn 1986 to harvest the beetroots and carrots. They were constantly coming across dead mice, and they joked among themselves that the mice would die out, then the beetles and worms, then the hares and wolves, and then us. People would be the last to die out. They began imagining a world without animals and birds. Without mice. For a time, there would be only people alive, all alone. There would not even be flies buzzing around. Those children were aged between twelve and fifteen. That is how they saw their future.

I talked to another girl. She went to a Young Pioneers' summer camp and made friends there with a boy. 'He was so nice,' she recalled. 'We spent all our time together.' But then his friends told him she was from Chernobyl, and he never came near her again. I even corresponded later with that young girl. 'When I think about my future now,' she wrote, 'I dream of completing school and going to some faraway place, so nobody knows where I come from. Somebody will fall in love with me there, and I will forget everything.' Yes, yes, write all this down, or it will slip people's memory and be lost. I only regret not writing everything down myself.

Another story. We came to a contaminated village. The children were playing ball by the school. It rolled into one of the flower beds. They stood there, walked around it, but were afraid to go and retrieve the ball. At first, I couldn't see the problem. I knew things theoretically, but I didn't live there. I wasn't constantly on the alert.

I came from the normal world. I walked over to the flower bed and immediately all the children shouted, 'No! No, mister. You mustn't!' In the past three years (this was in 1989), they had got used to the idea that you mustn't sit on the grass, or pick flowers, or climb a tree. When we took them abroad and said, 'Go for a walk in the forest. Go down to the river. Have a swim, sunbathe!' you should have seen how hesitant they were to go into the water, to stroke the grass. But then, afterwards . . . there was so much joy! They could dive in the river again, they could lie on the sand. They were forever walking around carrying bunches of wild flowers, and weaving them into circlets. What am I thinking right now? I am thinking that of course we can take them abroad and get treatment for them, but how are we to give them back the world they knew? How can we give them back their past? Or their future?

There is a question we cannot escape: who are we Russians? Until we answer it, nothing will change. What does life mean to us? What does freedom mean to us? We seem capable only of dreaming of freedom. We could have become free, but it didn't happen. We missed the boat again. For seventy years, we were building Communism, and today we are building capitalism. We used to worship Marx, and now we worship the dollar. History has passed us by. When you think about Chernobyl, you come back to the big question: who are we? What insights have we gained into ourselves? Into the world we inhabit? In our military museums, and we have more of those than we have museums of art, you find collections of old machine guns, bayonets, hand grenades, and out in the courtyard you see tanks and grenade launchers. Children are taken there on school trips and told: 'This is war. This is what war is like.' But actually, nowadays, it's completely different. On 26 April 1986, we faced war again; and that war is not over.

As for us . . . Who are we?

Gennady Grushevoy, member of the Belarusian Parliament, chairman of the Children of Chernobyl Foundation

Monologue on our having long ago come down from the trees
but not yet having come up with a way of making them
grow into wheels

'Do sit down. A bit closer. But I'll be perfectly frank, I don't like journalists, and they give me a hard time.'

'Really? Why is that?'

'Haven't you heard? Nobody got round to warning you? That explains why you're here, in my office. I am deeply suspect. That seems to be the consensus among your journalist colleagues. Everybody is shouting that it's not possible to live on that land, but I say it is. We need to learn how to live on it. We need to have courage. Should we seal off the contaminated area, put barbed wire round a third of our republic, abandon it, run away? After all, there's plenty more land in Russia. No! On the one hand, our civilization is antibiological. Nature's worst enemy may be man but, on the other, he is a creator. He can transform the world. You've only to look at the Eiffel Tower, or a spacecraft. But progress calls for sacrifices, and the more we progress, the more victims it calls for. No less than war. That's clear now.

'Pollution of the atmosphere, contamination of the soil, holes in the ozone layer . . . The earth's climate is changing. We're horrified, but we can't blame knowledge as such – or consider it a crime. Who is to blame for Chernobyl, the reactor or man? There are no two ways about it: man. He serviced it incompetently. There were outrageous errors, and the disaster was the sum of those errors. We won't delve into technical matters, but it's a fact. That's the conclusion reached by hundreds of commissions and experts. The largest man-made disaster in history. Our losses have been astronomical. The material losses we can more or less calculate, but what about the non-material damage? Chernobyl has blighted our imagination, our future. We are running scared of the future. But if that's the best we can manage, why did we bother coming down from the trees? Or perhaps we should have come up with a way of making them grow into wheels by themselves and cut out the middleman? In terms of casualties, it isn't the Chernobyl disaster that occupies the top spot in our world, but the motor car. Why is no one demanding a ban on the manufacture of cars? It's far safer to ride a bicycle or a donkey. Or in a cart.

'This is where they fall silent, my opponents.

'They ask me, "How do you feel about the children there drinking radioactive milk and eating radioactive berries?" I feel bad about it. Awful, actually! But I don't overlook the fact that those children have mummies and daddies, and that we have a government, and that it's their job to think about this. One thing I do get angry about, is when people who have forgotten, or never knew, Mendeleyev's periodic table try telling us how to live our lives. To intimidate us. Our Russian people have always lived in fear of war and revolution. That blood-drenched vampire, that Devil incarnate, Joseph Stalin . . . and now it's Chernobyl. And we wonder why people here are the way they are. Why aren't they free? Why are they so afraid of freedom? It's just that they are more used to living under a tsar: a father of his people. It makes not the least difference whether he's called the "general secretary" or the "president" . . .

'But I'm not a politician, I'm a scientist. All my life, I have been thinking about land and studying soil. The soil is a substance as mysterious as blood. We think we know all there is to know about it, and yet there is still some mystery. We divide, not into those in favour of living in this region and those against, but into scientists and non-scientists. If you succumb to appendicitis and need an operation, who do you to turn to? A surgeon, of course, not enthusiastic political activists. You'll take the advice of the expert. Well, I'm not a politician, but I ask myself, "What else do we have in Belarus besides soil, water and timber? Do we have an abundance of oil? Or diamonds?" We have nothing of that sort, so we should cherish what we have. Restore it. Yes, of course, other people sympathize. A lot of people in the world want to help us, but we can hardly live on Western handouts for the rest of time, relying on the contents of someone else's wallet. Everyone who wanted to has already left, leaving behind only people who want to live, not die, after Chernobyl. This is where they belong.'

'What are you proposing? How should people live here?'

'We can treat sick human beings, and we can treat sick soil too. We need to work, and think. We need to aspire to climb, perhaps only a little way at a time, to move forward. But instead, what are we doing? With our monstrous Slavic indolence, we prefer to believe more in miracles than in the possibility of actually doing something creative with our own two hands. Look at nature. We have to learn from her. Nature is working, cleansing herself, helping us. She behaves more rationally than man. She is striving to restore the original balance, the eternal order of things.

'I was summoned to the Communist Party's Provincial Executive Committee.

' "It's an uncommon situation. You will appreciate the problem, Slava Konstantinovna. We don't know who to believe. Dozens of scientists are saying one thing, and you're saying the opposite. Have you heard of this celebrated witch, Paraska? We decided to invite her here, and she has undertaken to lower the background gamma radiation over the course of the summer."

'You think that's funny? But these were senior people I was talking to. This Paraska had already signed contracts with several farms and been paid a lot of money. It was one of the fads we went through. Total eclipse of common sense. Generalized hysteria. Do you remember? Thousands, millions of people glued to their television sets, and sorcerers (who called themselves "psychics"), Chumak and Kashpirovsky, "energizing" water. My colleagues, people with degrees in the sciences, were filling up three-litre jars of water and moving them close to their screens. They drank it, they washed with it. It was supposed to have healing properties. The sorcerers performed in stadiums and drew audiences a popular singer like Alla Pugacheva could only dream of. People walked there, went by bus, crawled on their hands and knees, with incredible faith! To be cured of all ills by the wave of a magic wand! What was it all about? It was the new Bolshevik utopia. The audience were wildly enthusiastic, their heads stuffed full of it. "Now," I thought, "they think the wizards will save us from Chernobyl."

'He asked me directly: "What do you think? Of course, we're all atheists, but people keep saying . . . And it's in the newspapers. Should we arrange for you to meet her?"

'So I met Paraska the witch. Where did she come from? Probably the Ukraine. For the past two years, she'd been travelling all over the place, lowering the background radiation level.

' "What is it you are planning to do?" I asked her.

' "I have these inner powers . . . I sense I can reduce the background gamma radiation . . ."

' "What do you need for that?"

' "A helicopter."

'I was hopping mad. Both with Paraska and with our officials, who were listening slack-jawed while she made complete fools of them.

' "Oh," I said. "I don't think we'll need a helicopter just yet. We'll bring some contaminated soil and scatter it on the ground. Say, half a cubic metre. And you can lower its background radiation."

'Which is precisely what we did. The soil was brought, and she commenced. She whispered something, spat. Drove away some evil spirits with her hands and . . . what? And nothing. Paraska is in jail now somewhere in Ukraine, for fraud. Another witch guaranteed to speed the decay of strontium and caesium over a hundred hectares of land. Where did these people come from? I believe they materialized out of our desperation for a miracle, our expectations. Their photographs, their interviews . . . Somebody, after all, was giving them entire columns in the newspapers, prime time on television. If you lose faith in reason, all sorts of fears take its place, like in the mind of a savage. It produces monsters.

'This is where they fall silent, my opponents.

'I remember only one top official phoning and saying, "I would like to come and see you at your institute, and for you to tell me what a curie is. What is a microroentgen? How does this microroentgen turn into a pulse, for example? I travel round the villages, get asked these questions and look a complete idiot. Like a backward schoolboy." I met just one of our leaders like that. Alexey Shakhnov. Write his name down. Most of our leaders wanted to know nothing about physics or mathematics. The whole lot of them graduated from the Higher Party School, where they were properly taught just one thing: Marxism, inspiring and mobilizing the masses. It is the mentality of commissars and it hasn't changed since the days of Budyonny's cavalry army in the Civil War. I remember one of the favourite aphorisms of Stalin's favourite commander: "I don't care who I'm hacking at, I just like the swish of my sword."

'As for my recommendations on how we can live on this land? I'm afraid you'll find this boring, like everybody else. No sensations, no fireworks. How many times have I given talks in front of reporters, said one thing and read something quite different the next day. Their readers must have been scared witless. One person claims to have seen poppy plantations in the Zone, and settlements of drug addicts; another has seen a cat with three tails; there was a portent in the heavens on the day of the disaster, etc., etc.

'Here are the programmes our institute has developed. Printed checklists for collective farms and the public at large. I can give you this to keep. Spread the word. (*Reads.*)

' "Memorandum for Collective Farms. What do we recommend? Learn to treat radiation like electricity, directing it into channels where it harmlessly bypasses human beings. This means *perestroika*, restructuring and adapting the way we farm. In place of milk and meat farming, switch to industrial crops that do not enter the food chain. For example, rapeseed. It can be pressed to give oil, including motor oil. It can be used as fuel in engines. Seeds and seedlings can be grown. Seeds are actually subjected to radiation in the laboratory to maintain purity. It's harmless to seeds. That's one approach. There is a second. If we do, nevertheless, produce meat, we have no conventional way of decontaminating natural grain, but there is a way round this. It can be fed to cattle and, passing through them, is subjected to 'biological decontamination'. Before slaughter, bull calves are moved for two to three months on to indoor feeding, with non-contaminated fodder. They are decontaminated."

'I think that's probably enough. You won't want me to give you a whole lecture. These are science-based ideas. I would even call it a survival philosophy.

'We also have a Memorandum for Private Farmers, but when I come to a village to read it to old people, they stamp their feet. They refuse to listen because they want to live the way their grandfathers and great-grandfathers did, their ancestors. They want to drink fresh milk, but you mustn't do that. You need to buy a separator and make curd cheese with it, or churn butter. Drain off the whey and bury it. They want to dry mushrooms. Fine, but steep them first. Submerge them in water in a trough overnight, and only then dry them. Although it's safer still not to eat them at all. The whole of France is crazy about champignon mushrooms, but they don't grow them outside; they grow them in greenhouses. Where are our greenhouses? The cottages in Belarus are made of wood. Since time immemorial, Belarusians have lived in the forest, but their homes need to be lined with brickwork. Brick screens effectively, it dissipates ionizing radiation twenty times better than wood. Every five years, it is essential to lime the vegetable plots around the house. Strontium and caesium are cunning. They bide their time. You mustn't use the manure from your cow as fertilizer. You need to buy mineral fertilizers.'

'To realize your plans, you are going to need a different country, different people and different officials. In Russia, old people's pensions are barely enough for buying bread and sugar, and you are telling them to buy in fertilizers, a cream separator.'

'I can answer that. What I'm doing at this moment is defending science. I'm arguing that it's not science which is to blame for Chernobyl but people, not the reactor but people. Political matters aren't my department. You're talking to the wrong person there.

'Oops! I nearly forgot . . . and I'd even made a note to remind myself to tell you . . . A young scientist came to us from Moscow. He really wanted to be part of the Chernobyl project. Yury Zhuchenko. He brought his pregnant wife with him, in her fifth month. Everybody raised an eyebrow. What was he doing that for? The local people were fleeing, but other people were coming in from outside. He was doing it because he is a true scientist. He wanted to prove that someone who knew what they were doing could perfectly well live here. Intelligent and disciplined, precisely the two qualities we least admire. We admire people who block the enemy machine-gun position with their bare chest, who impetuously rush headlong forward bearing the flaming torch. But as for steeping mushrooms, pouring off the first water when the potatoes boil, taking vitamins regularly, bringing berries to the laboratory for testing, burying ashes in the ground . . . I went to Germany and saw every German there carefully sorting out their waste: bottles of clear glass in this container, green glass in that one. The lid of a milk carton went into the container for plastic, while the carton itself went into the container for paper. Camera batteries had a place of their own, compostable waste another. They really focus on it. I can't imagine a Russian behaving like that: white glass here, brown glass there. He would consider it too boring, simply beneath his dignity. God knows, his place is to make the rivers of Siberia run backwards, that sort of thing. The boundless Russian soul taking an axe to every problem . . . If we are to survive, we need to change.

'But that is not my problem. It is for you to solve. A matter of cultural attitudes, mentality, our whole way of life.

'This is where they fall silent, my opponents.' (*She is pensive.*)

'I would love to believe that soon the Chernobyl power station will be shut down, demolished, and that the area around it will be turned into a lush green lawn.'

Slava Konstantinovna Firsakova, doctor of agricultural sciences

Monologue by a capped well

The spring thaw meant I barely made it to the old farmstead. Our battered police jeep finally stalled, fortunately when we were already near the grand house, which was surrounded by spreading oaks and maples. I had come to see Maria Fedotovna Velichko, a singer and storyteller famed in the lowlands of Polesye.

I met her sons in the courtyard and we introduced ourselves. The elder was Matvey, a teacher, the younger, Andrey, was an engineer. They talked to me cheerfully, and everyone was clearly stirred up about their imminent move.

'The guest arrives as the lady of the house departs. We're taking my mother into town. We're waiting for the car. What book are you writing?'

'It's about Chernobyl.'

'It's interesting to be remembering Chernobyl today. I keep an eye on what gets written in the papers. There haven't been many books yet, though. As a teacher, I need to know about it. No one tells us how to discuss it with our children. I'm not that interested in physics – I teach literature – and the kinds of question I find fascinating are why Academician Legasov, who was one of those managing the aftermath of the accident, went back to Moscow and shot himself. Why the chief engineer of the atomic power station went out of his mind. Alpha particles, beta particles, caesium, strontium all decay, degrade, disperse . . . but what about human beings?'

'Well, I'm for progress! For science! None of us would dream of giving up the electric light bulb. Fear has been commercialized. Fear of Chernobyl is on sale because that's all we have to sell in the international markets. It's our new product. We sell our suffering.'

'They've evacuated hundreds of villages, tens of thousands of people. It's the great Atlantis of the peasantry. It's been scattered all over the former Soviet Union; it'll be impossible ever to put it back together again. It can't be saved. We've lost a whole world and there will never be another like it. Ask our mother.'

To my regret, this unexpected conversation, so earnestly begun, did not continue. There was urgent work to be done. I understood: these people were about to leave their family home forever.

But then the mistress of the house appeared in the doorway. She hugged and kissed me as if I were a member of the family.

'Daughter! Two winters I spent here on my own, with no one able to get through to see me but the animals. A fox skipped in one time. So surprised he was to see me! In winter, the days are long, and the nights no less so, like life. I could sing my songs to you about that, and weave you fairy tales. Life is boring for the old, and talking is what we do best. Time was, the students would come to see me from the big city, and record me on their tape recorders, but that was long since . . . before Chernobyl.

'What stories can I tell you? If there's even time for that . . . The other day I told my fortune from the water, and it said I would take a journey. Our very roots are being torn from their native soil. Our grandfathers and their fathers before them lived in these parts. They appeared here in the forests and replaced each other, century after century, but now such times have come that cares drive us from this land of ours. Such misfortune I find in none of the old tales. Really I don't know. Ah, but there . . .

'I'll remember for you, though, daughter, the way we told our fortunes when we were maids. I'll remember the good times and the laughter, the way my life here began. Life with my mother and father was so light-hearted until I was seventeen, and then it was time to think of a swain for me, to conjure my intended, to *gukat'*, as we say in these parts. In the summer we told fortunes from the water, and in winter from the smoke, and which way the chimney smoke blew was the direction you would leave in when you married. I loved to foretell the future from the water, the river . . . Water is what there was first on the earth, the water knows everything and can give you guidance. We would float candles on it, and pour wax. If the candle floated, then love was close; if it sank, for this year you would stay a maid and love would not find you. What was your lot? Where would you find happiness? We had so many ways of fortune-telling . . . We would take a mirror and sit in the bathhouse at night, and if someone appeared in the mirror, you must put it on the table at once or the Devil would pop out. The Devil likes to come through a mirror, from that other place . . . We would tell the future from shadows, burning paper over a glass of water and reading the shadow on the wall. If a cross appeared, death would come, but the dome of a church portended a wedding. One person would cry and another laugh, whatever was fated.

'At night, you would take off one of your shoes and put it under the pillow. In the night, your intended would come and take off the other one, and you needed to look at him and remember his face. Someone else came to me, though, not my Andrey, someone tall, his complexion fair, while my Andrey was short, his eyebrows black, and he was always laughing: "Oh, maiden mine, my lady be." (*Laughs*) We lived sixty happy years together. Brought three children into the world. Now he's no more. Our sons carried him to the grave. Before he died, he kissed me one last time: "Oh, lady mine, you'll now be all alone . . ." What do I know? You live a long life, forget how you lived, and even forget love. Ah, but there . . . It's all God's will. And when I was a maid, we would slip a comb under the pillow, and loosen our hair and sleep like that. And your intended would come to you in a dream and ask you for water to drink or to give to his horse . . .

'We would sprinkle poppy seeds around the well, in a circle, and in the evening gather round and shout down it, "Fortune, oo-oo-oo! Fortune, go-o-oh!" The echo would come back, and from the sound you would guess what was coming to whom. I wanted to go back just now to the well, to learn my future, although there is little enough left to me now. A few crumbs, a few dry grains. But the soldiers have sealed off all the wells. Hammered down planks over them, dead wells, capped. All that's left is an iron pump column near the collective-farm office. There was a wise woman in the village. She told fortunes too. Anyway, she went off to her daughter in the city. Sacks, two potato sacks full of herbs she took with her. I suppose it's God's will. Ah, but there . . . Some old clay pots she used for making her infusions. White sackcloth . . . Who needs that in the city? They just sit there, watching a television set or reading books. It's just us here, like birds, reading the land, the herbs and the trees. If the land in the spring took a long time to come through, if the snow didn't melt, you could expect drought in the summer. If the moon was weak and dark, new animals would not be born. If the cranes left early, the frosts would be severe. (*Swaying gently in time with her words.*)

'My sons are good men and their wives are loving. And my grandchildren. But who are you to talk to when you're out in the city? The place is strange to me. Empty for the heart. How can you reminisce among strangers? I loved going into the forest, we lived off it, there was always a group of us. You were among people.

You're not allowed into it now. Police are there, keeping watch on the radiation.

'Two years . . . God's will! Two years, my boys have been urging me, "Ma, come and live with us in the city." And finally they've got their way, in the end . . . But it is so lovely here, the forest all around, the lakes. The lakes so pure, with water nymphs. Old people told us that girls who died young lived on as water nymphs. Clothes were left for them, women's shifts, on the bushes. On bushes and hung on lines in the spring corn. They would come out of the water and run through the corn. Do you believe me? There was a time people believed all this. They listened. There was no television then, nobody had thought of it. (*Laughs.*) Ah, but there . . . Such beauty in this land! We lived here, but that will not be the lot of our children. No . . . I love this season of the year. The sun has climbed high in the sky, the birds have returned. Winter's tedium is over. There's no going out of the house of an evening, wild boar charging through the village as if they're in the forest. I sorted through the potatoes. Was going to plant up the onions. A body needs to be doing something, you can't just sit back and wait for death. That way he'll never come for you.

'And I remember too, daughter, the house sprite. He's been living in my house a long, long time, although I don't know exactly where, but he comes out from under the stove. Dressed all in black and with his black cap, and the buttons on his suit shining bright. No body to him, but he walks. I thought one time it was my old gentleman come back to visit me. Ah, well . . . But no . . . It was my kindly house sprite. I live alone, no one to talk to, so at night I tell him all about my day: "I came out early, before the sun was up, and there I stood and marvelled at the land. Rejoicing. So happy in my heart." But now we must be off, and I must forever part with my home. On Palm Sunday, I always picked the willow. We lacked a priest, so I would go down to the river and bless it there myself. I'd put it at the gate, bring it into the house and make it pretty. I'd stick it in the walls and doors, the ceiling, up under the roof. I'd say the words as I was going round: "Willow, willow take care of my cow. Make the corn grow strong and the apple trees fruit. May the chickens hatch and the geese lay well." That's how you have to go round and say the words, on and on.

'In the old days, we greeted the spring so merrily, playing and singing. We began on the first day, when the women let the cows out into the meadow. You have to ward off the witches. So they don't ruin the cows, don't steal their milk, so that when the cows come back home they haven't already been milked and have fear in their eyes. You need to remember all this, it may yet all come back again. It's all written down in the church books. When we still had a priest serving, he used to read that. Life can come to an end and then start all over again. Listen some more. Not many people still remember this, there's not so many left can tell it to you. Before the first herd goes out, you need to spread a white tablecloth on the road and let the herd run over it, and only then the herd-boys. And as they go, they should say these words: "Wicked witch, now gnaw at stones, now gnaw the earth . . . And you, my cows, pass safely over meadows and marshes, fearing none, unruly men or savage beasts." In spring, it's not only the grass comes out of the ground but everything else too. All sorts of evil. It hides away in dark places in your house, in the corners. In the cowshed, where it is warm. It'll come creeping into the yard from the lake, spread itself over the morning dew. A body needs to find protection. It's good to bury the soil from an anthill next to the gate, but the best thing is to bury an old lock by the gate. To lock the teeth of all the evil spirits, their lips. And the land? It needs not only to be worked by the plough and the harrow but also to be protected. From the evil spirits. You need to walk around your field twice and keep saying, "I sow the land, I scatter the seed, I wait for the harvest to be good, and may the mice not eat my wheat."

'What else can I give you to remember? In the spring, you also need to bow to pay respect to the stork, which we call the *busel-busko*. To thank it for flying back to its old nest. The *busel* protects you from fires and brings you babies. You call to it, "Klyo-klyo-klyo . . . *Busko*, come to us! Come to us!" And young people just married also call, "Klyo-klyo-klyo . . . May we love each other and be happy in love. And may our babies grow supple as the willow."

'At Easter, everybody painted eggs . . . Red eggs, blue, yellow . . . But if someone had died in your house, you painted just one black egg. In mourning. For sadness. But a red one was for love; a blue for long life. There now! Like me. I live and live. I know everything now: what will happen in spring and summer, autumn and winter. And for some reason, I am alive, looking at the world. Can't say I'm unhappy, daughter. Oh, and here's something else you should know. Put a red egg in water at Easter, let it lie there, and then wash your face. Your face will be pretty, pure. If you want to dream of someone dear who has died, go to their grave and roll an egg on the ground there: "Oh, mother dear, come to me. I need to share my unhappiness with you." And you tell her everything. Your life. If your husband is being unkind, she will tell you what to do. Before you roll the egg, hold it in your hands. Close your eyes and think. Don't be afraid of graves. It's only frightening when a dead person is being taken there. Close the windows and doors so death cannot fly in. He always wears white and carries a scythe. I haven't seen it myself, but people have told me, who have seen it. You mustn't let it see you. (*Laughs.*)

'If I'm going to a grave, I take two eggs, one red and one black. One in the colour of mourning. I'll sit down near my husband; there's a photograph of him on the monument, not young, not old, a nice photograph. "It's me, Andrey. Let's talk." I tell him all the news, and somebody calls me. A voice comes to me from some-where. "Hello there, lady mine." After I've visited Andrey, I go to my poor daughter. She died when she was only forty, the cancer got her. It didn't matter where we took her, nothing helped. She went young to her grave. So pretty . . . In the next world, they need all sorts too, old and young, pretty and ugly. Even little ones. But who's calling them all to the other side? I mean, what tales can they tell about this world? I don't understand, I just don't understand. But then, even clever people don't understand, professors in the city. Perhaps the priest in church knows the answer. When I see him, I'll ask. Ah, but there . . . What I say to my daughter is this: "Oh, my little daughter, my pretty one! What little birds will you fly with when you come back to me from far away? Will you come with the nightingales, or the cuckoos? What direction will you come from?" And so I sing to her and wait. Perhaps she'll appear to me, or give me a sign. Only you mustn't stay at the graves after dusk. At five o'clock, it's time to leave. The sun should still be high in the sky, and when it starts to move downwards, it's time to say goodbye. They want to be alone there. Like us. Just the same. The dead have their own life, just as we do. I don't know, but that's what I believe. That's what I think. Otherwise . . . Something else I ought to tell you: when a person is dying and suffering for a long time, and there are a lot of people in the house, everybody should go outside, to leave him on his own. Even his ma and pa should go out, and his children.

'Since dawn today, I've been walking round the yard and the vegetable plot, reminiscing. Those sons of mine have grown up to be good men, sturdy as oak. I've known happiness, but not much. I've worked all my life. How many potatoes alone have passed through these hands? Lugging them around. Ploughing, sowing . . . (*She repeats.*) Ploughing, sowing . . . And now . . . I'll bring out the garden sieve with all the seeds. I have seeds left – beans, sunflowers, beetroots – and I'll just scatter them over the bare earth and hope they live. And I'll shake the flowers over the yard . . . *Kvetochki* is our word for flowers . . . Do you know the scent of cosmos flowers on an autumn night? Especially before rain, they have such a fragrance. And sweet peas . . . But with the times as they are it's very unwise to touch the seeds. Scatter them over the ground and they will grow, be vigorous, but it's not for us humans. These times of ours . . . God has given us a sign . . . That day this damnable Chernobyl happened, I dreamed of bees, lots and lots of bees. Flying somewhere, flying. Swarm after swarm. Bees foretell fire. The earth will burn . . . God gave a sign to remind us: man is only a visitor on this earth, this is not his home, he is only visiting. We are visitors here . . .' (*She cries.*)

'Ma!' *one of her sons calls.* 'Ma! The truck's here . . .'

Monologue about longing for a role and a narrative

Dozens of books have been written, films made, commentaries provided. Yet this event is bigger than us, bigger than any commentator . . .

I heard, or read somewhere, that the problem we face with Chernobyl is primarily a problem of self-knowledge. I agreed. That fitted in with how I felt. I'm still waiting for some clever person to explain everything to me, to lay it all out clearly, in the same way that I'm being enlightened about Stalin, Lenin and Bolshevism. Or hearing endless propaganda about the market. The free market! We were brought up in a world where no Chernobyls existed, but now we find ourselves living with it.

Really, I'm a professional rocket engineer, an expert on missile fuel. I worked for the government in Baikonur. The Cosmos and Intercosmos programmes were a major part of my life. Conquer the heavens! Conquer the Arctic! Conquer the virgin lands! Conquer space! The whole Soviet world broke free of the earth, flew into space with Gagarin. All of us! I'm still in love with him! A wonderful Russian man with a wonderful smile! Even his death seemed somehow orchestrated. Dreams of soaring, flying, freedom. The longing to break free. That was such a marvellous time! For family reasons, I had to get transferred to Belarus and finished my army service here. When I got here, I immersed myself in this whole Chernobyl business. It altered my emotions. It was impossible to imagine anything comparable, although I'd always been dealing with the latest technology, space technology . . . It's difficult for the moment to find the words . . . It beggars the imagination. Something . . . (*Becomes pensive.*) A moment ago I thought I had caught the sense . . . a moment ago. You feel an urge to be philosophical. It doesn't matter who you speak to about Chernobyl, everybody feels that urge to philosophize.

But I'd do better telling you about my work. We're busy with so many things! We're building a church, a Chernobyl Church of the Icon of the Mother of God, the Seeker of the Perishing. We collect donations and visit the sick and dying. We're writing a chronicle, creating a museum. At one time, I thought I couldn't do it, that it was not in my heart to work in a place like this. I was given my first assignment: 'Here is a sum of money. Share it between thirty-five families, thirty-five widows whose husbands have died.' All of the men had been involved in the clean-up operation. I needed to do this fairly, but how? One widow had a little girl who was ill; another widow had two children; another woman was ill herself and renting an apartment; yet another had four children. I woke up in the night, wondering how I could avoid short-changing anyone. I thought and counted, counted and thought. Imagine it. In the end, I couldn't come up with anything, and we split the money equally among everyone on the list.

My real pet project is the museum. The Museum of Chernobyl. (*Silence.*) Only sometimes I get the feeling this is not a museum so much as a funeral director's office. As if I'm serving in a burial detail! This morning, before I had time to take my coat off, the door opened and a woman was there, sobbing. Well, not so much sobbing as yelling: 'Take his medal and all his certificates of merit! Take all his benefits – just give me back my husband!' She carried on shouting for ages, then left me his medal and the certificates. So now they will be displayed in a case in the museum. People will look at them . . . but no one but me heard what she shouted. Only I, when I'm arranging these exhibits, will remember.

Right now, Colonel Yaroshuk is dying. He's an expert in chemical dosimetry. He was a big, robust man, but now he lies paralysed. His wife has to turn him like a pillow, she feeds him with a spoon. He has kidney stones that need to be broken up, but we don't have the money to pay for the operation. We're beggars, existing on whatever people donate. The state, meanwhile, behaves like a con-man. It has abandoned these people; but when he dies, they will name some street, or school, or military unit after him. But only after he is dead. Colonel Yaroshuk. A man who walked through the radiation zone mapping the contamination hotspots. He was literally used as a biological robot, but even though he was aware of this, he moved out from right next to the walls of the atomic power station itself. On foot, carrying his radiation measuring instruments. If he detected a hotspot, he went along its edge so it could be accurately mapped.

And what about the soldiers working up on the reactor roof? In all, some 210 military units, about 340,000 troops, were brought in to clean up in the aftermath of the accident. The most lethal work was done by those cleaning the roof. They were issued lead aprons, but the radiation was coming up from below and they were not protected from that direction. They wore synthetic leather army boots and were exposed each day to intense radiation on the roof for one and a half to two minutes. Afterwards, they were discharged from the army with a certificate of commendation and a hundred-rouble bonus. They disappeared into the vast expanses of the Soviet homeland. On the roof, they were scraping up fuel and graphite from the reactor, lumps of concrete and reinforcing steel. They had twenty to thirty seconds to load a litter, and the same amount of time again to tip the waste down from the roof. These specialist litters weigh forty kilograms unloaded, so you can imagine the strain of wearing a lead apron, a mask, manoeuvring the litters and working at breakneck speed. A museum in Kiev has a mock-up of a piece of graphite the size of an army cap. The description states that, if it were real graphite, it would weigh sixteen kilos. That's how heavy, how dense it is. Radio-controlled handling equipment often malfunctioned because its electronic circuitry was burned out by the radiation. The most serviceable robots were soldiers. They were nicknamed 'green robots', after the colour of their army uniforms. Three thousand six hundred soldiers crossed the roof of the wrecked reactor. They slept on the ground. They all relate that, in the early days, straw was spread on the ground in the tents. The straw was taken from hayricks beside the reactor.

They were young lads. They're dying now too, but they know that, but for what they did . . . These are still people from a particular culture. A culture of superhuman feats and sacrificial victims.

There was a moment when there was a real risk of a nuclear explosion, and it was essential to drain the ground water beneath the reactor so it wouldn't be reached by a molten mix of uranium and graphite which, coming in contact with the water, would achieve critical mass. The power of the resultant explosion would have been three to five megatons. Not only would Kiev and Minsk have been wiped out, but an enormous area of Europe would have been made uninhabitable. A catastrophe on a European scale. The situation required volunteers to dive into the water and open the latch on the drainage valve. They were promised a car, an apartment, a dacha, and a pension for their families to the end of their days. Volunteers came forward! The boys dived, repeatedly, and managed to open the latch. They were given 7,000 roubles to share between them, and the promised cars and apartments were quietly forgotten about. Needless to say, they hadn't agreed to make those dives because of the promised rewards. That was the least part of their motivation. Our people are not that naive. They knew the value of those promises. (*Very upset.*)

Those people are no longer with us. There are only documents in our museum, names; but what if they hadn't done that? Our readiness to sacrifice ourselves . . . No one else comes close.

I had an argument about this with someone. He was trying to persuade me this was because we put a very low value on life. A kind of Asiatic fatalism. The person who makes the sacrifice has no sense of himself as a unique, irreplaceable human being. It is a longing for a role to play. Until this moment, he has been an actor with no lines, an extra. He has had no role, been no more than part of the background. Suddenly, he is a star. A longing for meaning. What is all our propaganda about? Our ideology? You are offered the choice of dying and acquiring meaning. They raise you up. They give you a role! It is worth dying because, afterwards, you will be immortal. He tried to persuade me of that, and gave examples, but I disagreed. Categorically! We are brought up to be soldiers. That is what we are taught. Constantly mobilized, constantly ready to undertake the impossible. My father was shocked when I wanted to go to a non-military university after school. 'I'm a career soldier, and you want to wear a civilian jacket? You need to defend the Fatherland!' He wouldn't speak to me for several months, until I applied to a military college. My father was a war veteran. He's dead now. Had practically no property, like everyone else of his generation. Nothing to bequeath: no house, no car, no land. What did he leave me? An officer's despatch case. He was given that just before the Finnish campaign, and now it holds his military awards. I also have a plastic bag containing 300 letters my father wrote from the front, beginning in 1941. My mother kept them. That is all he left, but I consider it a priceless legacy!

Now do you understand how I envisage our museum? Over there is a jar of Chernobyl soil, only a handful. Here is a miner's helmet, also from there. Peasant tools from the Zone. We can't let the radiation monitoring technicians in here. There's background radiation all right! But everything here must be authentic. No replicas! People have to be able to trust us, and they will only trust what is genuine, because there is just too much deceit about Chernobyl. That's how it was, and still is. You can use the atom not only for military and peaceful purposes, but also for selfish ones. Chernobyl has spawned countless charitable foundations and business interests.

Since you're writing this book, you need to view our unique videos. We're assembling them piece by piece. There is in effect no chronicle of Chernobyl; there was a ban on filming, everything was classified. If anybody did manage to record anything, the 'appropriate authorities' promptly confiscated the material and returned the tapes demagnetized. We have no documentary material about how people were evacuated or livestock was moved out. There must be no filming of a disaster, only of heroism! There have, nevertheless, been albums of Chernobyl photographs published; but how many times did film and television crews have their cameras smashed and how often were they dragged around various institutions to explain themselves? To talk honestly about Chernobyl took a lot of courage, and still does. Take it from me! But you need to see these images. The faces of the first firemen, as black as graphite. And their eyes, the eyes of people who already know they are not long for this world. One fragment shows the feet of a woman who went the morning after the accident to tend her vegetable plot next to the atomic power station. She walked on grass covered with dew. Her legs look like a sieve, peppered with holes right up to her knees. That is something you have to see if you are going to write such a book.

I come home and can't lift up my young son. I need to drink fifty or, preferably, a hundred mils of vodka before I take my boy in my arms.

A whole section in the museum is devoted to the helicopter crews. Colonel Vodolazhsky, a Hero of Russia, is buried in the soil of Belarus, in the village of Zhukov Lug. When he was subjected to a grossly excessive dose of radiation, he should have left. He should have been evacuated immediately, but instead he stayed and trained thirty-three other crews. He himself made 120 sorties and dropped 200 to 300 tonnes of ballast. He was making four or five flights a day, at an altitude of 300 metres above the reactor. The cabin temperature could reach sixty degrees. And what happened down there when they did drop the sandbags? Can you imagine the furnace? The level of radioactivity reached 1,800 roentgens per hour. The pilots were sick while they were flying. In order to aim accurately and hit the target, which was a volcanic crater, they stuck their heads out of the cabin and looked down. There was no other way. At meetings of the government commission dealing with the disaster, they reported matter-of-factly, 'We shall have to expend two or three lives on this . . . This operation will cost one life.' Straightforward and factual.

Colonel Vodolazhsky died. In their records of the radiation to which individuals were exposed above the reactor, the doctors wrote that he had been subjected to seven rem. The true figure was 600!

And what of the 400 miners who day and night dug under the reactor? They needed to carve out a tunnel that could be filled with liquid nitrogen to freeze the ground. Otherwise, the reactor would have sunk down into the ground water. There were miners from Moscow, Kiev, Dnepropetrovsk. I have never read a word about them and yet, naked, in temperatures of fifty degrees, they pushed mine carts in front of them on all fours. Down there, the radiation level was also in the hundreds of roentgens.

Today, they are dying; but what if they had not done what

they did? I consider them the heroes, not the casualties of a war which supposedly never happened. They call it 'an accident', 'a disaster', but it was a war. Our Chernobyl monuments resemble war memorials.

There are things that we do not believe in talking about, but you need to know them because of the kind of book you are writing. For those who worked at the reactor or in close proximity to it, what was most seriously affected – and this is very similar to the problems of those who work with missiles – was the genito-urinary system. Their masculinity. But Slavs just do not talk about these things. It's unacceptable. I once accompanied an English journalist who had prepared some interesting questions on this very topic. He wanted to investigate the human dimension of the problem. When it's all over, what happens to the human being when he goes back home, to his everyday life, to his sex life? He could find no one prepared to talk openly about it. For instance, he asked to meet the helicopter crews, to talk man-to-man. They duly came, some already retired at thirty-five or forty. One was brought along who had a broken leg caused by senile osteoporosis, because exposure to radiation causes bones to become brittle. The Englishman asked them how they were getting on in their families, with their young wives? The helicopter crews fell silent. They had come to talk about how they had flown five sorties a day, and here someone was asking them about their wives? About . . . He decided to try talking to them individually, in private. They all replied that their health was fine, the state valued them, and they had loving families. Not one of them would speak frankly. They left, and I could see he was distraught. 'Now you see,' he said to me, 'why nobody trusts you. You deceive even yourselves.' The meeting had taken place in a café, and they had been served by two pretty waitresses who were now clearing the tables. He asked them, 'Would you mind answering a few questions for me?' Those two girls gave him all the information he needed to know.

Journalist: 'Are you planning to get married?'

Waitress: 'Yes, but not here. We all dream of marrying a foreigner, so as to have healthy babies.'

Then he asked more boldly: 'Well, do you have a partner at present? How is everything? Do they satisfy you? You understand what I mean?'

Waitress: 'Look, you've just been interviewing those lads. (*They laughed.*) Helicopter crews. They're not far short of two metres tall, medals rattling on their chests. They're great at speaking from platforms, just not for going to bed with.'

Can you imagine it? He photographed the girls, and repeated what he had said to me: 'Now you see why nobody trusts you. You deceive even yourselves.'

I accompanied him into the Zone. It's a well-known statistic that there are 800 burial sites around Chernobyl. He was expecting some sort of amazing engineering structures, but they were just ordinary pits. They were filled with trees from the 'red forest' that was cut down in a 150 hectare area around the reactor. In the first two days after the accident, pine trees turned red and then russet.

There were thousands of tonnes of iron and steel, pipes, work clothes, concrete structures. He showed me an illustration from an English magazine, panoramic, from the air. Thousands of tractors, aircraft, fire engines and ambulances. The largest burial site was said to be next to the reactor. He wanted to photograph it now, ten years on, and had been promised a lot of money for the image. So there we were, being sent from one senior official to another. One said they needed a location from us, another that we needed a permit. We were just getting the run-around, until it dawned on me that this burial site did not exist. There no longer was a site in reality, only in reports. The machinery had long ago been looted and taken off to markets, to collective farms or people's homes for spare parts. It was all gone. The Englishman could not understand that. He could not believe it. When I told him the truth, he simply could not believe it! Now, when I'm reading even the most outspoken article, I don't trust it. At the back of my mind, I always have the thought that this may all be a lie as well, or a lot of fairy tales. Recalling the tragedy has become routine, a cliché. (*Hesitates.*) A horror story! (*Long pause.*)

I bring everything here, cart it all to the museum, but sometimes I think I should just forget it. Run away! It's too much.

I had a conversation with a young priest . . .

We were standing at the fresh grave of Sergeant Sasha Goncharov, one of the men who had been up on the reactor roof. Snow, strong winds, atrocious weather. The priest was conducting the funeral service, reciting the prayer, bareheaded. After the funeral, I said to him, 'You seemed not to be feeling the cold?' 'I don't,' he said. 'At times like that, I feel all-powerful. None of the other church services give me the kind of energy I receive from funerals.' I remembered that: the words of a man who is always in the vicinity of death.

I've often asked foreign journalists who come here – and many have come quite a few times – why they do it, why they apply to enter the Zone. It would be foolish to imagine they are only in it for the money or to advance their careers. 'We like being here with you,' they admit. 'It gives us a charge of energy.' Imagine that. It is a surprising answer, isn't it? For them, evidently, Slavs – the way we feel, the world we live in – are something they haven't come across elsewhere. The enigmatic Russian soul . . . We ourselves enjoy arguing about this over a drink in the kitchen. A friend of mine once said, 'Some day, we'll have everything. We'll have forgotten how to suffer. Then who will find us interesting?' I can't forget his words . . . but I've yet to work out what other people like about us. Is it ourselves, or that we are something to write about? That we help them to gain understanding?

Why are we always so fixated on death?

Chernobyl. There is not going to be a different world for us. At first, when they took the feet from us, we gave open expression to our pain; but now we live with the realization that there is no other world for us, and we have nowhere else to go. The sense that we are now forever fated to live on the soil of Chernobyl is something new. A lost generation returns from the war . . . Do you remember Remarque? But it is a perplexed generation that lives with Chernobyl. We are dismayed. The only thing that has not changed here is human suffering. That is our only currency – non-convertible, of course!

I come home after everything the day has thrown at me. My wife hears me out, but then quietly says: 'I love you, but I will not give up my son to you. I won't give him up to anyone, not to Chernobyl, not to Chechnya. Not to anyone!' She is already living with that fear.

Sergey Vasilyevich Sobolev, vice-chairman of the
Chernobyl Shield Association of Belarus

The Folk Choir

Klavdia Grigoryevna Barsuk, wife of a clean-up worker; Tamara Vasilyevna Belookaya, doctor; Yekaterina Fyodorovna Bobrova, resettled from Pripyat; Andrey Burtys, journalist; Ivan Naumovich Vergeychik, paediatrician; Yelena Ilyinichna Voronko, resident of Bragin; Svetlana Govor, wife of a clean-up worker; Natalia Maximovna Goncharenko, resettled; Tamara Ilyinichna Dubikovskaya, resident of Narovlya; Albert Nikolaevich Zaritsky, doctor; Alexandra Ivanovna Kravtsova, doctor; Eleonora Ivanovna Ladutenko, radiologist; Irina Yuryevna Lukashevich, midwife; Antonina Maximovna Larivonchik, resettled; Anatoly Ivanovich Polishchuk, hydrometeorologist; Maria Yakovlevna Savelyeva, mother; Nina Khantsevich, wife of a clean-up worker.

It's a long time since I saw a woman happy to be pregnant, or a happy mother . . .

A mother has just given birth. She comes to, and calls, 'Doctor, show me it! Bring it!' She touches the little head, the forehead, the little body. She counts the fingers and toes, checking. She wants to be quite sure: 'Doctor, have I had a normal baby? Is everything all right?' They bring him for her to feed. She is afraid. 'I live quite near Chernobyl. I was out in the black rain when it fell . . .'

They tell me what they have dreamed: one gave birth to a baby calf with eight legs; another to a puppy with the head of a hedgehog. Such bizarre dreams. Women didn't have dreams like these before. I never heard of it.

I've been a midwife for thirty years.

★

I've lived my whole life among words, and with words . . .

I teach language and literature in a school. I think it was in early June, when the exams were on. The headmistress suddenly assembled us and announced, 'You must all bring spades tomorrow.' We were given to understand we were to remove the top layer of contaminated soil around the school buildings, after which soldiers would come and cover everything with tarmac. Someone asked, 'What protection is being provided? Will they be bringing special suits and respirators?' We were told there would be nothing. 'Just bring your spades and dig.' Only two young teachers refused. The rest went and dug. People were not happy, but at the same time they had a sense of having done their duty. We have that in us: we ought to go where it is difficult and dangerous and be ready to defend our Motherland. What else have I been teaching the children? Precisely that. You have to go into battle, rush into the line of fire, defend, sacrifice. The literature I was teaching was not about life: it was about war and death: Sholokhov, Serafimovich, Furmanov, Fadeyev, Boris Polevoy . . . Only those two young teachers refused, but they belong to a new generation. They're different people.

We dug out the soil from morning till evening. When we were going home, it seemed strange that the city shops were open: women were buying stockings and perfume. We were already attuned to military thinking. It all seemed much more natural when suddenly there were queues for bread, salt and matches. Everybody rushed to dry out bread for storing. We washed the floors five or six times a day, sealed up the windows. We listened constantly to the radio. This way of behaving seemed familiar to me, even though I was born after the war. I tried to analyse my feelings and was amazed how quickly I reoriented my whole way of thinking. In some unfathomable way, I found I was remembering the experience of the war. I could picture myself abandoning our home and going off with the children, deciding what things we would take, what I would write to my mother. Even though ordinary, peacetime life was continuing all around, and they were showing comedy films on television.

Our memory was prompting us. We have always lived in horror. We're good at it. Horror is our natural habitat.

Our nation is unrivalled at that.

I wasn't involved in the war, but this reminded me of it . . .

The soldiers came into the villages and evacuated people. The streets were clogged with military vehicles: armoured personnel carriers, trucks covered with green tarpaulins, even tanks. People had to leave their homes with soldiers present. It was very oppressive, especially for those who had been through the war. At first, as Belarusians, they blamed the Russians. It was all their fault: it was their atomic power station. Later, they blamed the Communists. My heart was pounding with an almost superstitious sense of dread.

They tricked us. They promised we would be back within three days. We left behind our home, our bathhouse, the carved wooden covering over our well, our old garden. The night before we left, I went out into the garden and saw the flowers had opened. In the morning, they had all wilted. My mother did not survive the upheaval. She died a year later. I have two dreams which recur. In the first, I see our empty house; in the second, there are red dahlias by our garden gate, and my mother is standing there, alive, and smiling.

People are always comparing it to the war. War, though, you can understand. My father told me about the war, and I've read books about it. But this? All that is left of our village is three graveyards: one has people lying in it, the old graveyard; the second has all the cats and dogs we left behind, which were shot; the third has our homes.

They buried even our houses.

I wander through my memories every day . . .

Through the same streets, past the same houses. Our little town was so peaceful. There were no big manufacturing works, only a factory making sweets. It was Sunday. I was lying sunbathing. My mother came running. 'My dear, Chernobyl has blown up. People are hiding in their houses and you're out here in the sun.' I just laughed. Chernobyl is forty kilometres away from Narovlya.

In the evening, a Lada stopped outside our house. A girl I knew and her husband came in. She was wearing a housecoat, and he was in a tracksuit and some old trainers. They had sneaked out of Pripyat through the forest, down country lanes. They had escaped. There were police on duty on the roads, army checkpoints, and nobody was being allowed to leave. The first thing she shouted at me was, 'We need to find milk and vodka, urgently!' She was wailing and wailing. 'We'd just bought new furniture and a new fridge. I'd made myself a fur coat. We left everything, just covered it in plastic sheeting. We haven't slept all night. Oh, what's going to happen? What's going to happen?' Her husband was trying to calm her down. He told us there were helicopters flying over the town, and army trucks driving through the streets, spraying some sort of foam. They were calling up men for the army for six months, as if there was a war. We sat for days watching the television and waiting for Gorbachev to speak. There was no word from the authorities.

It was only after the May Day celebrations were over that Gorbachev appeared and said there was nothing to worry about, comrades, everything was under control. There had been a fire, just an ordinary fire. Nothing that unusual. The people living there were getting on with their work.

And we believed him.

Such scenes. I was afraid to sleep at night, to close my eyes . . .

They drove all the cattle from the evacuated villages to us in the district centre, to collection points. Maddened cows, sheep and pigs were running through the streets. Anyone who felt like it could just catch them. Trucks from the meat processing works drove to Kalinovichi station with carcases to be loaded and sent to Moscow, but Moscow refused to accept them. The wagons were already crypts and were sent right back to us. Whole trains at a time. The carcases were buried here. At night the smell of rotten meat was everywhere. 'Is this what nuclear war smells like?' I wondered. I thought war should smell of smoke.

In those first days, they evacuated our children at night, so not too many people would see. They tried to conceal the disaster, to hide it away. But, of course, people found out what was happening. They brought churns of milk to our buses for the journey, and baked buns for us.

Like during the war. What else can you compare it to?

A meeting at the Provincial Executive Committee. A war footing . . .

Everyone was waiting for the civil defence director to speak, because if anyone did know anything about radiation, it was only the fragments they remembered from the tenth-grade physics text-book. He came on to the stage and started telling us what was written in books and manuals about nuclear war: if a soldier is exposed to fifty roentgens of radiation he should be withdrawn from combat . . . He talked about how to build shelters, how to use a gas mask, the radius of the shock wave. But this was not Hiroshima or Nagasaki, everything was different. We were already into guesswork.

We flew a helicopter into the contaminated zone. The instruction manual specified: no underwear, cotton overalls like a chef (but with a protective film), mitts, a gauze mask. Festooned with all sorts of appliances, we descended out of the sky near a village, and there little children were splashing about in the sand like sparrows. Pebbles or twigs in their mouths. Little kids with no trousers, bare bottomed. And our orders were not to talk to people at all, so as not to cause panic.

Now I have to live with that.

Suddenly, there were television broadcasts . . .

One of the narrative tropes: a village woman has milked a cow. She pours milk into a jar, a reporter with a military radiation meter comes and sweeps it over the jar. There, see? Absolutely normal, and the reactor is only ten kilometres away. Footage of the River Pripyat, people swimming, sunbathing. In the distance, we can see the reactor with smoke above it. Commentary: 'Voices in the West are trying to sow panic, spreading outright slander about the accident.' Again the radiation counter is brought out, held over a bowl of fish soup, then over a bar of chocolate, then over doughnuts on sale at an outdoor kiosk. This was deliberate deception. The radiation meters our army had at that time were not designed for testing food. All they measured was background radiation.

The sheer volume of lies in our minds associated with Chernobyl bears comparison only with the situation at the outbreak of war in 1941 under Stalin.

I wanted the baby to be a token of our love . . .

We were expecting our first baby. My husband wanted a boy and I wanted a girl. The doctors urged me to have an abortion. 'Your husband was in Chernobyl for a long time.' He's a truck driver, and he was called up to go there in the early days. He was transporting sand and concrete. I wouldn't believe them. I didn't want to. I had read in books that love conquers all. Even death.

My little baby was stillborn, and lacking two fingers. A girl. I cried. 'If she could at least have had all her pretty little fingers. She was a girl after all.'

Nobody understood what had happened . . .

I rang the enlistment office. We medics are all army reservists. I offered my help. I don't remember the man's name, but he was a major. He told me, 'We only need young people.' I tried to change his mind. 'In the first place, young doctors are not trained, and in the second, they are more at risk. A young body is more sensitive to the effects of radiation.' His response was, 'We have orders to enlist young people.'

I remember patients' wounds began to heal poorly. Also that first radioactive rain, after which the puddles turned yellow. They turn yellow in the sun. I always find that colour unsettling now. On the one hand, the mind was simply not prepared for anything like that; on the other, we had always been told we were the best, the most amazing, and we were living in the world's greatest country. My husband is an engineer, he has a degree; but he seriously tried to persuade me it was an act of terrorism. Enemy sabotage. That's the way we thought, that's how we had been brought up. But I remembered travelling on the train with a technician who told me about the construction of the Smolensk atomic power station: how much cement and sand, how many planks and nails, found their way from the building site to nearby villages. In return for a bribe or a bottle of vodka.

In the villages and factories, people from the district committees of the Communist Party travelled around, meeting people. Yet not one of them was capable of giving an answer if they were asked what decontamination was, how children could be protected, or what the coefficients were for radionuclides finding their way into the food chain. Neither could they if asked about alpha, beta and gamma particles, nor about radiobiology, ionizing radiation, let alone isotopes. For them, that was all something from another planet. They gave lectures about the heroism of Soviet people, symbols of military courage, and the wiles of Western intelligence services.

I took the floor at a Party meeting and asked: 'Where are the professionals? The physicists? The radiologists?' They threatened to expel me from the Party.

There were a lot of inexplicable deaths . . . unexpected . . .

My sister had heart trouble. When she heard about Chernobyl, she had a premonition: 'You will survive this, but I won't.' She died a few months later. The doctors offered no explanation; and yet, with her diagnosis, she should have had many more years to live.

People talked about milk appearing in the breasts of old women, as if they had just had a baby. The medical term for it is 'relactation', but for peasants it was a punishment from God. This happened to an old lady who lived alone, without a husband or children. She went mad, 'lost her link to God', as they say there. She would wander through the village, rocking something in her arms. She would pick up a piece of wood or a children's ball, wrap her shawl round it, and sing it lullabies.

*

I'm frightened of living on this land . . .

They gave me a radiation meter, but what am I supposed to do with it? I wash the linen, I get it really clean, and the meter goes off. I make meals, bake a cake, it goes off. I make a bed, it goes off. What use is it to me? I feed the children and weep. 'Why are you crying, Mum?'

I have two children, both boys. I'm in and out of hospitals with them all the time, seeing doctors. You wouldn't know if the older one was a boy or a girl. His little head is quite bald. I've taken him to professors, and to the wise women too. Whisperers, witches. He's the littlest in his class. Can't run or play. If somebody hit him by accident, he would bleed. He could die. He's got a blood disease. I won't even name it. I stay in the hospital with him and think, 'He's going to die.' But then I see I mustn't think like that or death might hear. I cry in the toilet. All the mothers do. Not in the wards, but in the toilets or the bathroom. I come back all cheerful:

'Your cheeks are rosy pink. You're getting better.'

'Mum, take me away from the hospital. I'll die here. Everybody here dies.'

Where am I to cry? In the toilet? There's a queue. Everyone is just like me.

For Radunitsa, the Day of Remembrance of the Departed . . .

We were allowed to visit the graveyard, the graves . . . But the police ordered us not to go back to our houses. They were flying above us in their helicopters. We managed at least to glimpse our homes from far away, and make the sign of the Cross over them.

I brought back a sprig of lilac from my village. I've had it here for a year now.

Let me tell you what our people, Soviet people, are like . . .

In the early years, the shops in the polluted districts were over-flowing with buckwheat, and Chinese stewed meat, and people were so pleased. They boasted, 'There's no way anyone will get us to move out now. We're on to a good thing here!' Contamination of the soil was uneven; the same collective farm could have 'clean' and 'dirty' fields. You got paid more for working on the polluted soil and everyone wanted to work there. They would refuse to work in the clean fields.

Recently, I had a visit from my brother who lives in the far east of Russia. He said, 'They're using you here as "black boxes". The people here are black boxes like the ones they have on aircraft. They record all the information about the flight, and if a plane crashes they look for its black box.'

We think we are living life like everyone else. We walk around, go to work, fall in love . . . But no! Actually, we are recording data for the future.

I'm a paediatrician . . .

Children are completely different from adults. For instance, they're not afraid of death. It's a concept they don't have. They know everything about themselves: the diagnosis, the names of all the procedures and drugs. They know more than their mothers. And their games? They chase each other through the wards shouting, 'I'm radiation, I'm radiation!' When they die, it seems to me they look surprised. Baffled.

They lie there looking so surprised.

The doctors have warned me my husband is going to die. He has leukaemia . . .

He fell ill after coming back from the Chernobyl Zone. Two months later. He had been sent there by his factory. He came in from the night shift and said:

'I leave in the morning.'

'What are you going to be doing there?'

'Working on the collective farm.'

They were raking hay in the fifteen-kilometre zone, harvesting the beetroots, digging up potatoes.

He came home, went to see his parents. He was helping his father plaster around a stove and fell over. We called for an ambulance and he was taken to hospital. He had a lethally low level of white blood cells. They sent him to Moscow.

He came home with one thought in his mind: 'I am going to die.' Most of the time he just said nothing. I tried to change his mind, begged him, but he didn't believe a word. Then I gave birth to his daughter, to make him believe. I don't try to interpret my dreams. In one, I'm being taken to the scaffold; in another, I'm dressed all in white. I don't read those books telling you what dreams mean. I wake up in the morning, look at him and wonder how I could live without him. I want our little girl to grow up a bit so she'll remember him. She's so small. She's just started walking. She runs to him and says, 'Pa-pa.' I force myself not to think about it.

If I had known what was going to happen, I would have shut the doors, stood blocking the front entrance, locked them all ten times over . . .

I've been living at the hospital with my son for two years now . . .

Little girls play with dolls in the hospital wards. The dolls can close their eyes, and that's how you know a doll is dead.

'Why do the dolls die?'

'Because they're our children, and our children aren't going to live. They'll be born and just die.'

My Artyom is seven, but he looks five.

When he shuts his eyes, I think he's fallen asleep. That's when I can cry. When he won't see.

But he reacts:

'Mum, am I dying now?'

He falls asleep and he's barely breathing. I get down on my knees before him, in front of his little bed.

'Artyom, little one, open your eyes. Say something.'

I think to myself: 'You're still warm, my sweet one.'

He opens his eyes. He falls asleep again. He's so still. As if he's died.

'Artyom, dearest, open your eyes . . .'

I'm trying not to let him die.

We recently celebrated New Year . . .

We prepared a wonderful meal, everything homemade: smoked

sausage, pork fat, meat, pickled cucumber. We only bought the bread from the shop. We even had our own home-distilled vodka. Our very own, as people joke, Chernobyl-style, with added caesium and strontium to spice it up. Where else would we get it from? The shelves in the village shops are empty, and if anything does appear, it's beyond our wages and pensions to buy it.

Our visitors arrived. Our lovely neighbours. Young people. One's a teacher, the other a mechanic on the collective farm, with his wife. We drank. We ate. We sang. Without anyone organizing it, we started singing revolutionary songs. Songs about the war. 'A red dawn bathes in tender sunshine all the ancient Kremlin walls', my favourite. We had a wonderful evening. Just like old times.

I wrote about it to my son. He's studying in the capital, a student. I got his reply: 'Mum, I could just imagine this picture on the soil of Chernobyl. Our house. The New Year tree sparkling, and everybody at the table singing songs about the revolution and the war, as if there had never been a Gulag, never been a Chernobyl disaster . . .'

I felt so afraid. Not for myself, but for my son. He has nowhere to come home to now.

3

Admiring Disaster

Monologue on something we did not know: death can look so pretty

In the early days, the main question was: who is to blame? We needed a culprit . . .

Later, as we found out more, we began to wonder what we should do. How to get out of this mess. Now, though, when we are resigned to the thought that this is not going to be over in a year or two, but will last for many generations, we have started rethinking the past, turning back the pages one by one.

It happened on a Friday night. The next morning, no one had any suspicions. I sent my son to school, my husband went to the barber's. I was making lunch. My husband came back very soon, to tell me, 'There's been a fire at the atomic plant. We've been ordered not to turn the radio off.' I forgot to say we lived in Pripyat, near the reactor. To this day, I can see the bright, raspberry red glow. The reactor seemed lit up from inside. It was an incredible colour. Not an ordinary fire, but a kind of shining. Very pretty. If you forget all the rest, it was very pretty. I'd never seen anything like it in the movies, there was just nothing comparable. In the evening, everyone came out on to their balconies; if they didn't have one, they went to their friends and neighbours. We were on the eighth floor and had a great view. About three kilometres as the crow flies. People brought out their children and lifted them up. 'Look! Don't forget this!' And these were people who worked at the reactor:

engineers, workmen. There were even physics teachers, standing in that black dust, chatting away. Breathing it in. Admiring the sight. Some people drove dozens of kilometres or cycled to see it. We had no idea death could look so pretty. Not that there was no smell; it was not a springtime or an autumn smell. Quite different. Not the smell of soil either. No, it gave you a tickle in your throat and made your eyes water. I couldn't sleep all night, and heard the neighbours stamping about upstairs. They were awake too, hauling something about, banging, maybe packing things up, maybe sealing the windows. I took citramonum for a headache. In the morning, when dawn broke, I looked around. I'm not making this up now. It's not something I thought afterwards, but at that moment I sensed something was not right, something had changed. For good. By eight in the morning, there were soldiers with gas masks in the streets. When we saw them and the army vehicles there, we were not afraid. Quite the opposite, we were reassured. Now the army was here to help, everything would be fine. We had no idea the peaceful atom could kill, that the whole town might never have woken again after that night. Beneath our windows, someone was laughing, and music was playing.

After lunch, there were announcements on the radio that we should start preparing to evacuate. We would be moved away for three days, washed down and checked. I can still hear the announcer saying, 'evacuation to nearby villages', 'do not take pets with you', 'assemble downstairs by your apartment block entrance'. They said children must be sure to take their schoolbooks. My husband, nonetheless, put our identity documents and wedding photos in a briefcase. The only thing I picked up was a gauze headscarf in case of bad weather.

From the very beginning, we had a sense that we people from Chernobyl were now outcasts. Other people were afraid of us. The bus we were travelling in stopped for the night at a village. Evacuees were sleeping on the floor in a school or at a club and there wasn't an inch of spare space. One woman invited us to stay with her. 'Come, I'll make up a bed for you. I feel sorry for that son of yours.' Another woman nearby pulled her away from us. 'You're crazy!' she said. 'They're infectious.' After we had been resettled in Mogilyov, my son went to school. He burst into the house after his first day, crying. He had been sat next to a girl, and she complained she didn't want to sit there because he was 'radiated' and if she sat next to him she might die. My son was in fourth grade and, unluckily, he was the only person from Chernobyl in the class. They were all afraid of him. They called him the 'Glow-worm', or the 'Chernobyl Hedgehog'. It frightened me that his childhood came to an end so abruptly.

When we were leaving Pripyat, we found army convoys coming the other way. Armoured vehicles. That really was frightening, bewildering and frightening. But I somehow had the feeling that all this was happening, not to me, but to someone else. It was a very odd feeling. I was crying, searching for food and somewhere to stay, hugging and comforting my son, but inside I had this – it wasn't even a thought, just a constant feeling – that I was a spectator. I was looking through a window and watching someone else. It was only after we got to Kiev that we were given money, but there was nothing to buy with it. Hundreds of thousands of people had been moved from their homes. Everything had been bought up, eaten. Many people had heart attacks or strokes right there, at railway stations or on the bus, My mother was the saving of me. In the course of a long life, she had been deprived of her home and everything she had earned. The first time was when she was purged in the 1930s. They confiscated everything: her cow, her horse, her house. The second time it was a fire, and all she was able to save was me, her little daughter, plucked from the flames. 'You just have to get through it,' she comforted me. 'The main thing is, we're still alive.'

I've remembered something else. We were on the bus, crying. A man on the front seat was roundly cursing his wife: 'You're such a fool! Everybody else has brought at least a few things with them, but we're carting three-litre jars around.' His wife had decided that, as they were going on the bus, she would drop off the empty preserving jars for her mother along the way. They had these huge bulging string bags next to them, and we were all tripping over them throughout the journey. They travelled with them all the way to Kiev.

I sing in a church choir. I read the Gospel. I go to church, because it is the only place you hear talk of eternal life. That comforts people. Nowhere else will you hear words like these, and I so want to. When we were being evacuated, if we came to a church, everybody entered. It was almost impossible to get in. Atheists and Communists, they all went.

I often dream I am walking with my son through a sunlit Pripyat, although now, of course, it is a ghost town. We are strolling along, admiring the roses. There were lots of roses in Pripyat, big flower beds full of them. It's a dream. All that life of ours is a dream now. I was so young then. My son was little. I was in love.

Time has passed, everything is just a memory now. Again, I feel I'm a spectator.

Nadezhda Petrovna Vygovskaya, resettled from Pripyat

Monologue on how easy it is to return to dust

I kept a diary . . .

I tried to keep those days in my memory. A lot of new sensations. Fear too, of course. We were barging into something as alien as Mars. I come from Kursk. In 1969, an atomic power station was built near us, in Kurchatov. People went there from Kursk to shop for food, for sausage. People in the nuclear industry were provided with the top category of goods. I remember there was a big pond where anglers fished, near the reactor. After Chernobyl, I often thought about that. It wouldn't be allowed nowadays.

Right. So I was handed a conscription notice and, as an obedient citizen, reported the same day to the army enlistment office. The commissar leafed through my file: 'I see you haven't attended any of our training camps. Well, we need chemists. How do you fancy twenty-five days in camp near Minsk?' I thought, 'Well, why not take a break from work and the family? Do some marching around in the fresh air.' On 22 June 1986, armed with my belongings, a mess tin and toothbrush, I showed up at 11:00 hours at the muster point. It struck me there were just too many of us for peacetime.

Memories flashed through my mind from war movies. And what a day to choose: 22 June, the anniversary of the German invasion in 1941! We were ordered to fall in, then to stand down, and so on until evening. They boarded us on to buses as darkness was falling. We were ordered, 'Anyone who's brought alcohol, drink it. We'll get on the train tonight and be at the unit in the morning. I want you all as fresh as cucumbers and without a lot of baggage.' That was clear enough. We whooped it up all night.

In the morning, we found our unit in the forest. They lined us up again and did a roll-call, issuing protective clothing. One set, a second, a third. 'Uh-oh,' I thought. 'This looks serious.' They also issued us with a greatcoat, cap, mattress, pillow – all winter kit. But it was summer, and we had been promised we would be released after twenty-five days. 'Oh, come on, guys,' the captain transporting us laughed. 'Twenty-five days? You're on your way to Chernobyl for six months.' Bewilderment. Anger. Then they started talking us round: anyone inside the twenty-kilometre limit gets double pay, inside ten kilometres, triple pay, and at the reactor itself, multiply by six. One guy started calculating that in six months' time he would be able to drive home in his own car. Another just wanted to get out, but that would be desertion. What was radiation? None of them had heard of it.

As it happened, I had shortly before taken a course in civil defence. All the information dated from thirty years earlier: fifty roentgens was a fatal dose. They taught us how to hit the ground so the shock wave from the atomic bomb passed over without touching us. We heard about irradiation and thermal heating. There was never a word about the fact that radioactive contamination of the locality is the most damaging factor. The regular army officers taking us to Chernobyl knew precious little about it, but one thing they did know was that you should drink as much vodka as possible, because that had some effect against radiation. We were stationed near Minsk for six days, and for six days we drank. I made a collection of the labels from the bottles. First we drank vodka, but then there were some odd beverages: Nitkhinol and various other window-cleaning fluids. As a chemist, I took a professional interest. After imbibing Nitkhinol, your legs were like jelly but your head was clear. You could order yourself, 'Stand up!' but you'd then fall over.

So there I was, a chemical engineer with a candidate of sciences degree, called away from my job as the laboratory manager of a large corporation, and what use was made of me? I was handed a shovel and that was virtually my only tool. A slogan was promptly born: 'Shovels against the atom!' Our protective equipment was respirators and gas masks, but nobody used them, because in heat of up to thirty degrees you would croak almost immediately. We signed for them as 'spare ammunition' and forgot about them.

Another detail. Our transport: we transferred from the buses to the train. There were forty-five seats in the carriage, and seventy of us. We took it in turns to sleep. I just remembered that. Well, anyway, what was Chernobyl like? Army equipment and soldiers. Field showers. A military environment. We were accommodated in tents for ten people. Some of the men had left children behind at home, one lad's wife was having a baby, another didn't even have a home. There was no whingeing. If this job had to be done, someone had to do it. The Motherland had called, commanded. That's the way we are.

Outside the tents, there were mountains of empty tin cans. Veritable Mont Blancs! Emergency supplies that had been stored somewhere in army depots. Judging by the labels, they had been stored for twenty or thirty years. In case of war. Tins that had contained stew, barley porridge or sprats. There were herds of cats. They were like flies. The villages had been evacuated, no people around. You would hear a garden gate creak and you'd turn round, expecting to see a person, but instead there would be a cat.

Contaminated topsoil was removed, loaded on to trucks and taken away to the burial sites. I imagined a burial site for contaminated materials would be some complicated technical structure, but it was an ordinary barrow-like mound. We lifted the surface soil and rolled it up, like a carpet. It was green turf with grass, flowers and roots, worms and spiders. What we were doing was insane. You can't strip away all the earth, taking out of it everything that is alive. If we had not got diabolically drunk every night, I doubt we could have kept going. The human mind just isn't that resilient. Hundreds of square metres of flayed, barren land. Houses, barns, trees, major roads, kindergartens, wells were left naked. In the middle of a desert, filled up with sand. When it was time to shave in the morning, you were afraid to look at your own face in the mirror, because all sorts of thoughts came to you, all sorts. It's hard to believe people have gone back there and that life has started up again. But we were changing slates, washing roofs. Everybody knew it was completely pointless. Thousands of people. But every morning, we got up and got on with it. Totally absurd! An illiterate old grandad came up to us: 'Forget it, young fellows, that's no work to be doing. Come home and have dinner with us.' The wind was blowing, dark clouds sailing in the sky. The reactor was not sealed. We would take off a layer, come back a week later, and we might as well have started all over again. There was nothing left to strip. The radioactive sand was drifting down. Only once did I see something that made sense, and that was when helicopters were spraying the ground with a special solution that formed a polymer membrane and stopped the loose soil from blowing about. I could see the point of that. But we just kept on digging.

The locals had been evacuated, but there were still a few old people in some villages. And then, just to be able to go into an ordinary cottage and sit down for a meal – the ritual of it – for even half an hour of normal human company . . . Although you mustn't eat anything: that was strictly prohibited. But you so wanted to sit down at that table, in an old cottage.

By the time we had finished, all that was left was burial mounds. They were supposed then to be faced with concrete slabs and fenced in behind barbed wire. They left behind the tipper trucks, jeeps and cranes that had been used, because metal absorbs and accumulates radiation. I heard, though, that it all subsequently disappeared off somewhere. Looted. I can believe it, because in our country anything is possible. There was an alarm one time: the radiation technicians checked out where the canteen was built and found the radiation there was higher than at the sites where we were working. By then, we'd been living there for two months. That's the kind of people we are. There were poles with planks nailed on to them at chest-height, and that was what they called a canteen. We ate standing up, washed out of a barrel, and the toilet was a long trench in an open field. We were battling an atomic reactor, armed with shovels.

After two months, we were beginning to think we were being had for suckers. We decided to protest. 'We're not kamikazes. We've spent two months here and that's enough. It's time we were relieved.' Major-General Antoshkin gave us a pep talk and was entirely frank: 'It's not in our interests to relieve you. We've given you one set of clothing, another, a third. You've learned how to do the work. It would be an expensive business, relieving you, and a lot of bother.' He put great emphasis on what heroes we were. Once a week, whoever had really excelled at dirt-digging was presented with a certificate of merit in front of all of us, on parade. Top Radioactive Earth-Shoveller of the USSR. How mad was that?

Empty villages with only cats and chickens living there. You go into a barn and it's full of eggs. They used to fry them. Soldiers are reckless. They'd catch a chicken, make a fire, get a bottle of moonshine . . . Every day, they got through a three-litre jar of vodka in our tent. Some would be battling it out at chess, someone would be strumming a guitar. People get used to anything. One would get drunk and go to bed, another might like to shout and yell. And fight. Two got drunk, drove off and crashed a vehicle. Had to be cut out of a tangle of iron with an oxyacetylene torch. I stayed sane by writing long letters home and keeping a diary. The head of the political section spotted me and started snooping, trying to find out where I kept it and what I was writing. He told one of the others in my tent to spy on me. The lad inquired, 'What are you scribbling?' I said, 'Oh, I've got my candidate's degree and now I'm working on a PhD.' He laughed and said, 'I'll tell the colonel that, but you'd better hide it somewhere.' They were good lads. I've already said there were no whingers, not one coward. Believe me: no one will ever beat us. Not ever! The officers never came out of their tents. Slobbing about in slippers. Getting drunk. To hell with the lot of them! We did the digging. Let them get new stars on their epaulettes. Good luck to them! That's the kind of people we are.

The radiation monitoring technicians were gods. Old people were always pestering them. 'Tell me, sonny, what's my radiation like?' One enterprising soldier had a bright idea. He took an ordinary stick and wrapped some wire round it. He knocked at one cottage and ran his stick over the wall. An old woman was soon after him: 'Sonny, what's that thing saying about my home?' 'Military secret, Granny.' 'You can tell me, sonny. I'll pour you a glass of moonshine.' 'Oh, okay!' He drank it and said, 'Everything's fine, Granny', and on he went.

Finally, about halfway through our time, we were issued dosimeters, little boxes with a crystal inside. Some of the soldiers worked out that it would be a good idea to take it away in the morning, leave it by a burial site, and pick it up again at the end of the day. The higher the radiation reading, the sooner they should be discharged. Or paid more. Some put it on their boot. There was a strap there you could hang it on, so it was close to the ground. Talk about theatre of the absurd! It was crazy! The sensors were not electrically charged, which was essential if they were to start measuring. In other words, those little packs of nonsense, those baubles were issued purely to con us. As psychotherapy. It turned out these silicon gadgets had been lying around in storage for fifty years or so. At the end of our stint, we had a notional number entered in everybody's army record book: the average dose of radiation multiplied by the number of days on site. The 'average dose' was measured at the tents we were living in, not where we were working.

Was it a story or was it true? A soldier rang his girlfriend. She was worried: 'What are you doing there?' He decided to boast: 'I've just come out from very near the reactor and washed my hands.' Immediately, the dialling tone. Conversation terminated. The KGB eavesdropping.

Two hours' rest. You lie down under a bush. The cherries are ripe, and so big and sweet . . . You give them a quick wipe and pop them in your mouth. Mulberries. I saw a mulberry bush for the first time.

When there was no work to do, they would march us about. Over the contaminated ground. Absurd! In the evening we watched movies, Indian ones, about love. Until three or four o'clock in the morning. If the cook overslept, our breakfast porridge was under-done. They brought newspapers. They were writing in them about what heroes we were. Volunteers! Heirs of Pavel Korchagin! They printed photos. If we had met up with that photographer . . .

There were international units stationed nearby. Tatars from Kazan. I saw their kangaroo court. One lad ran the gauntlet. If he stopped or tried to run to one side, they gave him a kicking. He had been climbing up on the roofs of houses to clean them, and they found a bag of stuff he had looted. The Lithuanians were in a separate camp. After a month, they mutinied and demanded to be sent home.

One time, we got an order to go immediately and wash down a house in an empty village. Absurd! 'Whatever for?' 'There's going to be a wedding there tomorrow.' The roof was hosed down, and the trees, and the ground was scraped off. The potato haulms in the garden and the grass in the yard were scythed down. There was a complete wasteland all around. The next day, the bride and groom were brought in, together with a busload of guests. Live music. They were a real bride and groom, not actors. They had been reset-tled and were now living somewhere else, but were persuaded to come here and be filmed for the historical record. The propaganda machine was working, the dream factory, preserving our myths: we can survive anywhere, even in a dead land.

Just before I was demobbed, I was called in by the commander: 'What have you been writing?' 'Letters to my young wife,' I replied. 'You just watch it', was his final order.

What has stayed in my memory from that time? Us digging and digging. I noted somewhere in the diary what I understood while I was there. In the very first days, I realized how easy it is to return to dust.

Ivan Nikolaevich Zhmykhov, chemical engineer

Monologue on the symbols and secrets of a great country

I remember it as a war . . .

By the end of May, about a month after the accident, we began receiving produce from the thirty-kilometre zone to test. Our institute was working round the clock, like in wartime. At that period, we were the only place in the republic with the professional staff and specialized equipment. They brought us the innards of domestic and wild animals. We tested milk. After the first tests, it became clear that what we were dealing with was not meat but radioactive waste. Herds in the Zone were being tended in shifts. The cowherds were brought in and then left. Milkmaids were brought in only for the duration of the milking. The dairy factories were fulfilling their Plans. We tested what they were producing. It was not milk, it was radioactive waste. For years afterwards, we used dry milk powder and cans of condensed and evaporated milk from the Rogachev Dairy in our lectures as a benchmark. At that time, though, it was all on sale in the shops, on all the food stalls.

When people read on the labels that milk was from Rogachev, they rejected it and the stocks piled up. Then jars suddenly started appearing without labels. I don't think the reason for that was any shortage of paper – people were being deceived deliberately. Deceived by the state. Every piece of information had become a secret, at precisely the time when short-lived elements were emitting maximum radiation and everything was 'glowing'. We were continually writing internal memoranda. Continually. But to say anything publicly about the results would see you stripped of your academic degree, and even your Party card. (*Becoming nervous.*) But it was not fear. Not because of fear, although we were fearful, of course . . . But because we were people of that time, citizens of our Soviet land. We believed in it. It was all to do with faith. Our faith . . . (*Lights a cigarette in his agitation.*) Believe me, it was not from fear . . . or not only from fear. I'm answering you honestly. For my own self-respect, I need to be honest now. I want . . .

That first trip to the Zone: in the forest, the background radiation was five or six times higher than in the open countryside or on the road. Everywhere, the readings were high. Tractors were being used. The peasants were digging their plots. In several villages, we checked the thyroid readings of adults and children: in 100 of them, the radiation was 200 or 300 times above the acceptable level. We had a woman in our group, a radiologist. She became hysterical when she saw children sitting in the sand, playing. Sailing their little boats in puddles of water. The shops were open and, as usual in our villages, clothing and food were alongside each other: suits, dresses, and next to them, sausage and margarine. They were lying there openly, not even covered with plastic sheeting. We took sausage and eggs and sampled them: they were not food, they were radioactive waste. A young woman was sitting on a bench by her house, breastfeeding. We tested her breast milk and it was radioactive. The Madonna of Chernobyl.

We asked, 'What are we to do about this?' We were told, 'Carry on testing and watch the television.' On television, Gorbachev was being reassuring. 'Emergency measures have been taken.' I believed him. I – an engineer with twenty years' experience, someone who knew the laws of physics. I knew every living thing needed to be evacuated from that area, at least temporarily. But we conscientiously carried on, making our measurements and watching the television. We were accustomed to believing. I belong to the post-war generation that grew up with that faith. Where did it come from? We had been victorious in a dreadful war. At that time, the whole world admired and respected us. That was really true! In the Cordilleras, the name of Stalin was carved into the rocks. What did that mean? It was a symbol! It meant we were a great country.

So there's the answer to your question of why we knew and said nothing. Why didn't we shout it from the rooftops? We reported the situation. I told you, we wrote internal reports. And we stayed silent and obeyed orders implicitly, because we were under Party discipline. I was a Communist. I don't remember any of our staff being afraid for their own skin and refusing to travel to the Zone. And that was not because they were afraid of losing their Party card, but because of their faith. Above all, a belief that we were living in a fine and just society that put people first. Man was the measure of all things. For many people, the collapse of that faith ended in a heart attack or suicide. A bullet in the heart, as with Academician Legasov. Because when you lose that faith, when you are marooned without faith, you are no longer part of something, but complicit in it, and you no longer have any justification. That is how I understand what he did.

A significant detail: every nuclear power plant in the former Soviet Union had a contingency plan in the safe for dealing with an accident. There was one standard plan. Secret. Without such a plan, permission would not be granted for the station to begin operating. It was devised, many years before the accident, on the basis of the Chernobyl power station: what was to be done and how. Who was responsible for what. Where they were to be. Everything down to the last detail. And then suddenly, at that very power station, there is a disaster. What are you to make of that? Was it coincidence? Mystical forces? If I was a believer . . . When you need to find meaning, you do get a religious feeling. I'm an engineer and I subscribe to a different faith. I defer to other symbols . . .

But what am I do now with my faith? What now? . . .

Marat Filippovich Kokhanov, former chief engineer of the Institute of Atomic Energy, Belarus Academy of Sciences

Monologue on the fact that terrible things in life happen unspectacularly and naturally

From the very beginning . . .

Somewhere something had happened. I didn't even catch the name, but it was somewhere far away from our Mogilyov. My brother came running from his school and told me all the children were being issued pills of some sort. Evidently something really had happened. Oh dear, oh dear! But that was all. We had a wonderful time on May Day. Out in the countryside, naturally. We came home late at night. The wind had blown the window in my room wide open. I remembered that later.

I worked at the Nature Conservation Inspectorate. We were expecting to receive instructions, but none came. We waited. There were almost no professionally trained people among the inspectorate's staff, especially among the top management: retired colonels, former Party workers, pensioners or officials who were under a cloud. If they slipped up somewhere else, they were exiled to our inspectorate. They sat there, shuffling papers. They woke up and started speaking out after our Belarusian writer Ales Adamovich really sounded the alarm in Moscow. How they hated him! It was unreal. This was where their children and grandchildren were living, but it was a writer, not them, who cried out to the world for help! You would have thought, if nothing else, the instinct for self-preservation would have kicked in; but at Party meetings and in the smoking rooms they were all ranting on about this wretched pen-pusher. Why couldn't people mind their own business? He had slipped the leash! There were orders from above! What about subordination! What did he know? He wasn't a physicist! We had a Central Committee! We had a general secretary! That was probably the first time I realized what the purges in 1937 were like. How it must have been.

At that period, my idea of a nuclear power station was quite idyllic. At school and in college, we were taught that these were fairy-tale 'factories making energy out of nothing', in which people in white coats sat and pushed buttons. Chernobyl exploded in minds which were completely unprepared, which had complete faith in technology. That was made worse by the total lack of information. There were mountains of papers stamped 'top secret', 'Classify information about the accident as secret', 'Information on outcomes of medical treatment to be classified', 'Information on extent of radiation poisoning of personnel involved in the clean-up to be classified'. Rumour was rife: someone had read in the newspapers, someone had heard somewhere, someone had been told . . . Libraries were stripped of everything that had been published on civil defence, which all turned out to have been complete garbage anyway. Some people listened to the 'voices' broadcast by the West, which were the only source of information at the time on which pills to take and how to optimize their effectiveness. More often, however, the reaction was that the enemy was gloating, but actually we were just fine. On 9 May, Victory Day, the ex-servicemen and women would parade, there would be a brass band playing. Even those trying to extinguish the reactor, it turned out, were dependent on rumours. Apparently picking up graphite with your bare hands was dangerous . . . apparently.

From somewhere, a madwoman appeared in the city. She went round the market saying, 'I've seen this radiation. It's as blue as blue can be, glistening . . .' People stopped buying milk and curd cheese. An old lady was standing there trying to sell milk, but nobody wanted it. 'Don't be afraid,' she wheedled, 'I don't take my cow out in the open, I bring her the grass myself.' If you drove out into the countryside, you would find models of animals sticking up along the roadside: a cow covered in plastic, grazing, and beside it a village woman, also wrapped in plastic. You didn't know whether to laugh or cry. They had started sending us out into the field for monitoring. I was sent to a timber camp. There had been no reduction in the quantity of timber coming in for processing. The Plan remained unaltered. We switched on our equipment at the depot and the radiation was sky-high. Next to the planks seemed more or less all right; but next to the finished brooms, it was off the scale. 'Where have these brooms come from?' 'Krasnopolye' (subsequently found to be the most contaminated area in the whole of our Mogilyov Province). 'That's the last batch left. All the rest have been sent out.' How could we track them down now they were all over the place?

There was something else I really didn't want to forget. Something very striking . . . Ah, yes. Chernobyl . . . Suddenly, we had this new, unfamiliar awareness that each of us had his or her own life. Until then, that hadn't seemed to matter. Now, though, people began thinking about what they were eating, and what they were feeding to their children. What was bad for your health, and what was safe. Should you move away or stay? Everybody had to decide for themselves. But how were we used to living? As part of the village, the commune, the factory or collective farm. We were Soviet people. I, for one, was entirely Soviet. When I was studying at college, I went off every summer with a Communist Detachment. There was a youth movement at the time called Student Communist Detachments. We worked, but our pay was transferred to some Latin American Communist party. Our detachment was supporting Uruguay.

We had changed. Everything had changed. It needs a very great effort to understand, to break away from what you are used to. I'm a biologist. My thesis was on the behaviour of wasps. I spent two months on an uninhabited island. I had my own personal wasps' nest there. They accepted me as a member of their family after a week of sizing me up. They didn't allow anyone closer than three metres, but within a week I was allowed to within ten centimetres. I fed them jam from a matchstick right in their nest. 'Do not destroy an anthill: it is a splendid alien life form', was a favourite saying of our lecturer. A wasps' nest is linked with everything in the forest, and I too gradually became part of the environment. A baby mouse would run up and sit on the edge of my trainers. He was wild, a forest animal, but he already saw me as part of the landscape. I had been sitting here yesterday, I was sitting here today and I would be sitting here again tomorrow.

After Chernobyl . . . In an exhibition of children's drawings, a stork walks over a black field in springtime. The caption: 'Nobody told the stork anything.' That was my feeling at the time. And I had work to do. Every day. We travelled round the province, collecting samples of water, samples of soil, and taking them to Minsk. Our girls grumbled, 'These are really hot cakes we are carrying.' There was no protection, no specialist clothing. You sat in the front, and behind you were 'glowing' samples. We wrote up instructions on how to dispose of radioactive soil. Burying soil in the soil, that's a new occupation for the human race. Nobody could understand what they were supposed to be doing. In our manual, the dumping of waste was supposed to follow a geological survey to ensure that the ground water was no less than four to six metres down, and that the disposal pit was shallow. Its sides and base were to be lined with plastic sheeting. But that was only what it said in the instructions. The reality was different. No geological survey. They stabbed a finger in the map and said, 'Dig there.' The operator digs. 'So how far down did you go?' 'The Devil only knows! When I saw water, I stopped.' They dumped it straight into the ground water.

They say we are God's people with a criminal government. I'll tell you afterwards what I think about our people, and about myself.

My most important assignment was in Krasnopolye District. As I mentioned, that was absolutely the worst affected area. To prevent radionuclides getting from the fields into the rivers, there were things that needed to be done in accordance with instructions: to plough double furrows, then leave a space, then again plough double furrows, and so on in the same pattern. I needed to drive along all the minor rivers, monitoring. I took the bus to the district centre; from there, of course, I needed transport. I went to see the chairman of the executive committee. He sat in his office, clutching his head in his hands: nobody had cancelled the Plan, nobody had changed the crop rotation. If this was the year for sowing peas, then peas they would sow, even though they knew peas, like all legumes, absorbed maximum amounts of radiation. For heaven's sake, in places there were readings of forty curies or more! He had no time to listen to me. The cooks and nurses had fled the kindergartens and the children were hungry. If anybody needed an operation, they had to be taken by ambulance to the neighbouring district, sixty kilometres away, and the road was as bumpy as a washboard. All the surgeons had fled. What transport was I imagining? What double furrows was I talking about? He had no time to listen. I headed to the army. Young lads working a six-month stint there. Now they're desperately ill. They assigned me an armoured personnel carrier and crew. In fact, not just that but a reconnaissance vehicle with a machine gun. I was very sad not to be photographed riding on the armour. More romanticism! The corporal in command kept in constant touch with his base: 'Falcon! Falcon! Mission proceeding.' On we drove, our roads, our forest, and us in this combat vehicle. Women were standing by their fences, crying. The last time they had seen this kind of armour was during the Second World War. They were afraid this was a new war.

The instructions specified that the tractors used for ploughing these furrows must have radiation shielding and a hermetically sealed cabin. I did see such a tractor. It actually did have an airtight cabin. There it stood, with the tractor driver lying on the grass taking a nap. 'Are you crazy? Has nobody warned you?' 'It's okay. I've covered my head with my jacket,' he replied. People had no understanding. They had been constantly readied to expect a nuclear war, but not a Chernobyl.

It was such a beautiful area. The trees were not recent plantings but the original, ancient forest. Meandering streams, their water tea-coloured but so, so limpid. Green grass. People calling to each other in the forest. But I knew it had all been poisoned, the mushrooms, the berries, the squirrels scampering among the nut bushes.

We met an old lady: 'My dears, is it safe for me to drink the milk from my cow?'

Our eyes downcast. We have orders to collect data and avoid talking to the local population. The corporal came to his senses first: 'Grandma, how old would you be?'

'Oh, already past eighty, maybe more. My papers all got burned during the war.'

'Well, go ahead. Drink it.'

I feel most sorry for the country people. They were the real victims, as innocent as children. Chernobyl was not something any peasant had invented. They had their own relationship with nature, a trusting, not predatory, attitude. Just like a hundred years ago, or a thousand, in accordance with God's providence. They couldn't understand what had happened. They wanted to have the same faith in scientists, in anyone who could read and write, as they had in the priest. And they were constantly being told: 'Everything is fine. No cause for alarm. Just remember to wash your hands before meals.' I came to see, not then but some years later, that we had all been complicit. In a crime. (*Falls silent.*)

Goods were pilfered from the Zone by the truckload. Everything that was sent there as aid, as gifts to the people living there: coffee, tinned stew, ham, oranges. Crates of the stuff, vanloads. At that time, food like that was simply not available elsewhere. The local traders did very nicely, as did every inspector, all those petty and mid-level bureaucrats. Human beings proved more despicable than I had realized. That was true of me too. I was despicable too. That's something I know now about myself. (*Becomes pensive.*) Of course, I have to admit this. It's important for me. One more example. In one collective farm, for instance, there might be five villages: three 'clean', two 'dirty'. They were all two to three kilometres apart. Two would be getting compensation, three wouldn't. In a 'clean' village, they would build an animal-breeding centre. 'We'll bring in clean fodder.' Well, where was that supposed to come from? The wind carries dust from one field to another. It's all just one farm. To build the centre requires permits and a commission to authorize them. I am a member of that commission, even though everyone knows we ought not to sign this project off . . . It's a crime! In the end, I found myself an excuse: the problem of uncontaminated fodder is no concern of a nature conservation inspector. I'm only a small person. What can I do about it?

Everybody found an excuse, an explanation. That is an experiment I have conducted on myself. I realize now that terrible things in life happen unspectacularly and naturally.

Zoya Danilovna Bruk, nature conservation inspector

Monologue on the observation that a Russian always wants to believe in something

Have you really not noticed that we don't talk about this even among ourselves? In several decades', several centuries' time, these years will be seen in terms of myth, the landscape peopled with folk tales and legends.

I'm afraid of rain. That's what Chernobyl means. I'm afraid of snow, of forests, of clouds. Of the wind . . . Yes! Where's it blowing from? What's it bringing? That's not an abstraction, not a rational consideration, but my personal feeling. Chernobyl is in my own home. It's in the being I most cherish: my son. He was born in spring 1986, and he is ill. Animals, even cockroaches, know when to give birth and to how many progeny. Human beings can't do that. Their Creator hasn't bestowed on them the gift of premonition. It was in the newspapers recently that in 1993, here in Belarus alone, women had 200,000 abortions. The main cause was Chernobyl. We are living with that fear. Nature has, as it were, pulled back, waiting. 'Woe is me! Whither has time gone?' Zarathustra might exclaim.

I've thought a lot about these things. Searching for meaning, an answer. Chernobyl was a disaster of the Russian mentality. Has that not struck you? I agree, of course, when people write that it wasn't just a reactor that exploded but the entire old system of values. But to me this can't be the full explanation.

I would like to talk about something first discussed by Pyotr Chaadayev in the late 1820s: our hostility to progress, our opposition to technology, our instinctive distrust of tools and instruments. Look at Europe. Since the Renaissance, it has professed an instrumental approach to the world. Intelligent, rational. Respect for the craftsman, the tools in his hands. There's a splendid story by Nikolai Leskov, 'Iron Will'. What is it about? The Russian character is all about hoping for the best and muddling through. That's the epitome of Russianness. The German character puts its trust in tools and machines. What about ours? On the one hand, there's an attempt to rein in and overcome chaos; on the other, there's our much-loved impulsiveness. Go anywhere you like, to Kizhi in the north of Russia, for example, and what will you hear, what will the guide proudly be boasting about? That the famous wooden church there was built using only an axe and that it doesn't have a single nail in it! Rather than build good roads, we want to put a horseshoe on a flea. The cartwheel may sink in the mud, but we get to hold the Firebird in our hands.

The second feature, I think . . . Yes! This is the price we've paid for our rapid industrialization after the October Revolution, for that leap forward. Again, in the West, there was a century of spinning and manufacturing. Man and machine moved forward and changed in tandem. There was time for a technological awareness, a technological way of thinking to develop. But in Russia? What does our peasant have in his farmyard, other than his own hands? To this day! An axe, a scythe, a knife – and that's all. That's what holds his world together. Oh, and a shovel. How does a Russian interact with a machine? By cursing it. Or taking a sledgehammer to it. Or giving it a kick. He doesn't like the machine; in fact, he hates and despises it. The truth is that he has no idea at all of the power he has in his hands.

I read somewhere that the operators of atomic power stations would often call the reactor the 'cooking pot', the 'samovar', the 'paraffin stove', the 'hob'. There's a certain jejune arrogance here: we'll fry eggs on the sun! There were a lot of village people among those working at the Chernobyl plant, a lot of rustics. During the day they were at an atomic reactor, and in the evening they were working their vegetable plots or visiting their parents in the next village, where they still plant potatoes with a spade and spread manure with a pitchfork. They bring in the harvest by hand too. Their thinking was switching between two eras, the Nuclear Age and the Stone Age. A person was constantly swinging like a pendulum. Imagine a railway laid by brilliant track engineers, and a train hurtling along it; but instead of train drivers in the cab, you have the drivers of horse-drawn carriages from yesteryear. Coachmen. Russia is fated to travel simultaneously in two cultures. Between the atom and the shovel. As for technological discipline . . . For Russians, discipline has always been associated with coercion: the stocks, chains. The people wanted to be spontaneous, liberated. The dream was not of freedom but of liberty. For us, discipline was a means of repression. Our ignorance has a peculiar quality, something close to oriental barbarism.

I'm a historian. Previously, though, I was very interested in linguistics, the philosophy of language. It's not simply a matter of us thinking language; language also thinks us. When I was eighteen, and perhaps even a little earlier, when I began reading samizdat and discovered Shalamov and Solzhenitsyn, I suddenly realized that my entire childhood, and the childhood of everyone in my street – and I grew up in an educated family (my grandfather was a priest, my father a professor at St Petersburg University) – had been suffused with a prison-camp mentality. All the vocabulary of my childhood was the language of those prisoners, the *zeks*. For us boys, that was entirely natural. You called your father 'the boss', your mother 'the momma'. 'For smartasses, there's a screw-action dick.' I learned that expression at the age of nine. Yes! Not a single civil word. Even our games, sayings and riddles came from the camps. Because the *zeks* didn't inhabit a separate world of prisons that were far away. It was all around us. 'Half the country imprisoning, half the country imprisoned,' as Anna Akhmatova said. I believe it was inevitable that this prison camp mindset was predestined to collide with culture, with civilization, with the Dubna synchrophasotron.

Of course, it's just a fact . . . We were brought up in a particular kind of Soviet paganism. Man was almighty, the crown of creation. He had the right to do whatever he pleased with the world. Ivan Michurin's phrase was much quoted: 'We cannot wait for the favours of nature; our mission is to take them from her.' The attempt to inculcate in the people qualities and attributes they did not possess. The dream of global revolution was an aspiration to remake human beings and the world around us. Remake everything! Yes! There's that renowned Bolshevik slogan: 'With an iron fist we shall herd the human race into happiness.' The psychology of a rapist. The materialism of a caveman. Defying history, defying nature. And it's still going on. One utopia collapses and another comes to take its place. Everyone has suddenly started talking about God. God and the market, in the same breath. Why didn't they go looking for him in the Gulag, in the dungeons of the Purges in 1937, at the Party meetings in 1948 which set out to smash 'cosmopolitanism', under Khrushchev when they were destroying churches? The present-day subtext of Russian God-seeking is evil and deceitful. They bomb the homes of the civilian population in Chechnya, trying to wipe out a small, proud nation, and then stand in a church holding candles. We can do nothing except by the sword. We use the Kalashnikov instead of words. They scrape the charred remains of Russian crews out of tanks in Grozny using shovels and pitchforks, whatever's left of them. And at the same time, we have the president and his generals praying. Russia watches all that on television.

What do we need? An answer to the question whether the Russian nation is capable of a wholesale review of its past in the same way that the Japanese managed after the Second World War. And the Germans. Do we have enough intellectual courage? They're saying nothing about that. All the talk is about the market, privatization vouchers and pay cheques. We are back once again at just trying to survive. All our energy goes on that. The soul is cast aside. The individual is isolated again. But in that case, what's the point of it all – of your book, of my sleepless nights – if our life is like the momentary flaring of a match? There can be different answers to that. One is primitive fatalism. But there can also be magnificent answers. A Russian always wants to believe in something: in the railway, in dissecting frogs (like Turgenev's Bazarov), in Byzantinism, in the atom . . . And now, in the market.

In Bulgakov's *Molière*, one of the characters says, 'All my life I have sinned. I was an actress.' This belief that art is sinful, that it is fundamentally depraved, peeping into someone else's life. And yet, like a serum from someone who is ill, it can vaccinate you with other people's experiences. Chernobyl is a subject for Dostoevsky. An attempt to justify the existence of man. Or perhaps everything's much simpler than that, and instead we should approach the world on tiptoe and stop at the doorway!

Contemplate this divine world with awe . . . and live our lives like that.

Alexander Revalsky, historian

Monologue about how defenceless a small life is in a time of greatness

Don't ask. I won't do it. I don't want to talk about it . . . (*Aloof silence.*)

No, I will talk to you, I want to understand. If you'll help me. Only don't feel sorry for me, don't try to console me. Please! Don't do that! No – to go through such suffering without there being any meaning to it, to have to rethink so much, is wrong. Impossible! (*Starts shouting.*) We're back again on the same old reservation, back again living in one big prison camp. The Chernobyl prison camp. They rant at their rallies, carry their slogans, write in their newspapers . . .

Chernobyl brought down an empire. It cured us of Communism. It cured us of feats of heroism no better than suicide, of terrifying ideas. I understand now. Those 'feats of heroism' are a concept the state invented. For people like me. But I've nothing more than that left, nothing else. I grew up surrounded by words like that and people like that. Everything has gone, that life has gone. What is there to hold on to? What can rescue me? To go through such suffering without there being any meaning to it is wrong! (*Silence.*) One thing I do know is that I'll never be happy again . . .

He came back from there. For a few years, he seemed to be living in a delirium. He told me all about it, everything. I remembered it all.

There was a red puddle in the middle of the village. The geese and ducks walked round it.

Soldiers, just boys, with their boots off, their clothes off, lying on the grass, sunbathing. 'Get up, you morons, or you'll die!' They laughed: 'Ha ha ha!'

Many people wanted to drive away from the villages in their cars. The cars were contaminated. They were ordered, 'Everybody out!' and their car was dumped in a special pit. People were standing there, crying. They would come back at night and secretly dig it out.

'Nina, I'm so glad we have two children.'

The doctors told me his heart was half as big again as it should be, and his kidneys, and his liver.

One night, he asked, 'Aren't you afraid of me?' He had started being afraid of intimacy. I didn't ask him about it. I understood him, with my heart. I wanted to ask you . . . I wanted to say . . . I often think . . . I sometimes feel so bad I don't want to know the answer. I hate remembering! I hate it! (*Starts shouting again.*) At one time . . . At one time, I envied heroes. Those people who had a part in great events, when everything was in the balance, conquering great summits. That was how we talked then, what we sang about. Such splendid songs. (*Sings.*) 'Little eagle, little eagle . . .' I can't even remember the words now. 'Fly higher than the sun . . .' Is that right? What wonderful words our songs had. I used to dream! I felt sad I wasn't born in 1917 or 1941. I think differently now. I don't want to live history, in historic times, where my short life is just so defenceless. Great events trample it underfoot without even noticing, without pausing. (*Pensive*.) After us, history is all that will be left. Chernobyl will be left . . . But my life? My love?

He told me all about it, everything. I remembered it all.

The pigeons, the sparrows . . . The storks. A stork would run and run across the field, trying to take off but not able to. A sparrow would scuttle over the ground, jumping and jumping, but not able to fly, not able to fly over the fence.

The people had gone and only their photos went on living in their homes.

They were driving through an abandoned village and saw a fairy-tale picture: an old man and woman sitting together on their porch with hedgehogs running around them. There were so many of them, like little chicks. It was quiet in the village with no people, as if in the depths of the forest. The hedgehogs were no longer afraid to come and ask for milk. The foxes would come, the old couple told them, and elk. One of the lads couldn't keep his mouth shut and said, 'Me, I'm a hunter!' 'No, no, young fellow!' the old people chided him. 'You mustn't hurt the animals! We are related to them now. All one big family.'

He knew he would die, that he was dying, and promised himself he would live only for love and friendship. I had two jobs because his pension was not enough to keep us, but he said, 'Let's sell the car. It's not new, but all the same we should be able to get something for it. Stay at home. I'll just look at you.' He invited friends. His parents came and lived with us for a long time. He had understood something. He had discovered something there about life which he hadn't known before. The words he spoke were different.

'Nina, I'm so glad we have two children. A boy and a girl.'

I would ask him, 'Did you think about me and the children? What did you think about while you were there?'

'I saw a little boy there. He was born two months after the explosion. His name was Anton, but everybody called him Atom.'

'Did you think . . .'

'You felt sorry for everyone and everything there. Even for a gnat or a sparrow. Let everything live. Let the flies fly, the wasps sting, the cockroaches scuttle . . .'

'Did . . .'

'Children draw pictures of Chernobyl. The trees in the pictures grow with their roots in the air. The water in the rivers is red or yellow. They cry while they are doing their drawings.'

His friend . . . His friend told me it was incredibly interesting there, a lot of fun. They recited poetry, sang songs with someone playing a guitar. The best engineers and scientists came. The elite of Moscow and Leningrad. Philosophizing. Alla Pugacheva came and sang for them, in the open air. 'If you don't fall asleep, boys, I'll sing for you till morning.' Heroes, she called them. His friend . . . was the first to die. He was dancing at his daughter's wedding, making everyone laugh with his stories. He lifted a glass to propose a toast and collapsed. Our men, all dying as if there was a war on, but this is peacetime. I don't want to go on! I don't want to remember . . . (*Closes her eyes, rocking gently to and fro.*) I don't want to talk. He died, and it was so dreadful, such a dark forest . . .

'Nina, I'm so glad we have two children. A boy and a girl. They will live on.'

(*She continues.*) What is it I want to understand? I don't know that myself . . . (*A barely perceptible smile.*) Another of his friends proposed to me. When we were at college, when we were students, he courted me, then married my friend, but they soon divorced. Something didn't work out between them.

He brought me a bunch of flowers: 'You will live like a queen.' He owns a shop, a fashionable apartment in town, a dacha in the country . . . I turned him down. He was offended. 'It's been five years now . . . Can you really not forget your hero? Ha ha . . . You're living with a monument . . .' (*She shouts.*) I kicked him out! I kicked him out! 'You're a silly fool! Go on, then, live on your teacher's salary, your hundred dollars.' And so I do. (*Calming down.*) Chernobyl filled my life and my heart grew bigger . . . but it aches. It's like a secret key. After suffering great pain, you talk, you find you speak well. I did . . . I only found that language when I really loved. And now. If I didn't believe he's in heaven, how could I have survived that?

He told me. I remembered it all . . . (*She speaks as if in a trance.*)

Clouds of dust . . . Tractors in the fields. Women with pitchforks. The radiation meter clicking away . . .

No people, and time moves differently . . . A long, long day, like when you were a child . . .

You weren't allowed to burn leaves. They had to be buried.

To go through such suffering without there being any meaning is wrong. (*Crying.*) Without those wonderful words we knew so well. Even without that medal they gave him. We've got it in the cupboard at home . . . Something he left us.

But one thing I do know is that I'll never be happy again.

Nina Prokhorovna Litvina, wife of a clean-up worker

Monologue on physics, with which we were all once in love

I'm just the man you need. You've come to the right place . . .

Since I was in my teens, I've made a habit of writing everything down. For example, when Stalin died, what was happening on the streets, what was reported in the newspapers. As for Chernobyl, I was keeping a record from the first day. I knew that, as time passed, much would be forgotten and lost forever. That's exactly what has happened. My friends, nuclear physicists who were in the centre of all the activity, have forgotten what they felt at the time, what they talked to me about; but I have it all written down.

That day, I came to work – I'm the laboratory director at the Institute of Atomic Energy of the Belarus Academy of Sciences. Our institute is in the countryside, in woodland. Wonderful weather! Spring. I opened the window. The air was clean and fresh. I was puzzled that none of the tits appeared that I had been feeding all through the winter. I would hang out pieces of sausage for them. Where had they gone?

Meanwhile, panic had broken out at our institute's reactor. The radiation measuring equipment was indicating increased radioactivity; on the air purifying filters, it had increased 200-fold. The dose rate by the institute's entrance was about three milliroentgens per hour. That was very serious. That level was the maximum permissible for working in hazardous areas for not more than six hours. Our first hypothesis was that the shell of one of the fuel rods in the core was leaking. We checked and found no abnormality. Alternatively, we thought that while the container was being transported from the radiochemical laboratory it had been struck so hard that the inner shell had been damaged and our territory contaminated. It would be quite some job to clean that mark off the tarmac! So what could it be? At this point, there was an announcement over the internal radio that staff were advised not to go outside the building. Suddenly, the area between the buildings was empty. Not one person to be seen. Very odd. Eerie.

The radiation monitoring technicians checked out my office and found the table, my clothes and the walls all 'glowing'. I stood up; I'd no desire to sit down on a chair. I washed my head over the sink and checked the meter again: that had had a beneficial effect. So was the problem with our institute? Was there a leak? How could we decontaminate the buses that transported us around the city? How could we decontaminate our staff? I needed to rack my brains. I was very proud of our reactor. I had studied it down to the last millimetre.

We phoned the nearby Ignalina atomic power station in Lithuania. Their instruments too were shrieking. They too were in a panic. We tried to call Chernobyl, but could get no reply from any telephone. By lunchtime the picture was becoming clearer. There was a radioactive cloud over the whole of Minsk. We identified it as iodine radioactivity, which indicated an accident at a reactor.

My first reaction was to phone home to warn my wife, but all our phones at the institute were tapped. Oh, this eternal fear hammered into us for decades! But they knew nothing about what was happening. After lessons at the Conservatory, my daughter would be walking about the city with her friends. Eating ice cream. Should I phone? But I could get in trouble. I might be banned from classified work. I decided I couldn't just do nothing. I picked up the phone.

'Listen carefully.'

'What are you on about?' my wife asked loudly.

'Speak softly. Close the windows, put all the food in plastic bags.

Put on rubber gloves and wipe everything you can with a damp cloth. Put the cloth in a bag too and put it well out of the way. Put any laundry drying on the balcony back in the washing machine. Don't buy any bread, and under no circumstances buy cakes in the street.'

'What's happened?'

'Shhh! Dissolve two drops of iodine in a glass of water. Wash your head . . .'

'What . . .' But I did not let my wife finish and hung up. I thought she would understand. She worked at the institute herself. And if some KGB operator was listening, he had probably written the life-saving recommendations down on a piece of paper for himself and his family.

At 15:30 hours, we established that there had been an accident at the Chernobyl reactor.

That evening, we returned to Minsk in the institute's bus. For that half hour, we said nothing, or talked about other matters. We were afraid to speak to each other about what had happened. Each of us had a Party card in his or her pocket.

There was a wet cloth lying outside my apartment door. My wife had understood. I went into the hallway, took off my suit and shirt and stripped to my underpants. I suddenly felt so angry. To hell with all this secretiveness! This fear! I took the city telephone directory, the address books of my daughter and wife, and began phoning everyone in turn. 'I work at the Institute of Atomic Energy. There's a radioactive cloud over Minsk.' I listed all the precautions they needed to take: wash your hair with household soap; close all your windows; wipe the floor every three or four hours with a wet cloth; clothes off the balcony, back in the wash. Take iodine. How to do that correctly. People's reaction was just to thank me. No questioning, no fear. I suspect they either didn't believe me, or were unable to grasp the scale of the incident. No one was afraid. An amazing reaction. Mind-boggling!

That evening, my friend phoned. A nuclear physicist, doctor of sciences. How carefree, how trusting we were! It's only now you see that. He called to say, among other things, that he was planning to go for the May Day holiday to his wife's parents in Gomel Province. That was a stone's throw from Chernobyl. He would be taking small children. 'What a brilliant decision!' I shouted. 'You're mad!' That was our sense of professionalism, and our sense of trust. I yelled at him. He probably doesn't remember now I saved his children's lives.

(*After a pause.*) We – I'm talking about all of us – haven't forgotten Chernobyl, we just haven't understood it. What do savages understand about lightning?

In a collection of essays by Ales Adamovich, there's his conversation with Academician Andrey Sakharov about the atomic bomb. 'Do you know what a marvellous smell of ozone there is after a nuclear explosion?' the father of the Soviet hydrogen bomb exclaimed. There's a romanticism in these words. My romanticism, the romanticism of my generation. Forgive me. I see the reaction on your face. You think this is delight at a global nightmare rather than at human genius. But it is only today that atomic energy has been humiliated and disgraced. My generation . . . In 1945, when the first atomic bomb was detonated, I was seventeen. I loved science fiction. I dreamed of flying to other planets and believed atomic energy would lift us into space. I enrolled at the Moscow Energy Institute, and learned that there was a top-secret department of high-energy physics. The 1950s, the 1960s . . . Nuclear physicists were the elite. Everybody was wildly excited about the future. Those studying humanities were pushed aside. In a three-kopeck piece, our school teacher had told us, there is enough energy to operate a power station. It took your breath away! I couldn't read enough of what Cyril Stanley Smith wrote about the invention of the atomic bomb, the conducting of tests and details of the first explosion. In the USSR, all this was kept secret. I read, I imagined. In 1962, *Nine Days in One Year*, a film featuring Soviet atomic scientists, was released. It was very popular. They earned large salaries and the secrecy added to the romanticism. The cult of physics! The golden age of physics! Even after it imploded at Chernobyl, how slow we were to part with that cult! Scientists were summoned. They arrived on a special flight, but many didn't even take razors with them, supposing they would be away for only a few hours. They were informed that there had been an explosion at an atomic power station, but all had faith in their physics. They were from the generation which shared that belief. The Age of Physics ended at Chernobyl.

You already have a different view of the world. I recently found in my favourite philosopher, Konstantin Leontyev, the idea that the results of the depraving of physics and chemistry would sooner or later oblige the cosmic intelligence to intervene in our earthly affairs. Those of us brought up in Stalin's time could not countenance the notion that any supernatural powers might exist. Or parallel worlds. I read the Bible later, and married the same woman twice. I left and then came back. We met again. Who is going to explain that miracle to me? Life is the most amazing thing! Enigmatic! Now I believe . . . in what? That the three-dimensional world is no longer big enough for modern man . . . Why is there so much interest nowadays in an alternative reality? In new knowledge? Man is breaking away from the ground . . . He is operating with different categories of time, and not just with the earth, but with different worlds. Apocalypse. The Nuclear Winter. This has all been described in Western art already. Depicted. Filmed. They've prepared themselves for the future. The explosion of a large number of nuclear weapons will cause an enormous conflagration. The atmosphere will be saturated with smoke. The sun's rays will be unable to reach the ground and that will set off a chain reaction of cold, colder, colder still. This secular vision of the end of the world has been around since the Industrial Revolution in the eighteenth century, but the atomic bomb won't disappear even when the last warhead has been decommissioned. The knowledge will remain.

You're not saying anything. And yet I am arguing with you the whole time. We have an argument going on between generations. Have you noticed? The story of the atom is not just about state secrets, a mystery or a curse. It is also the story of our youth, our future. Our religion . . . But now? Now I also believe that the world has another ruler, and that we, with our guns and spaceships, are like children. I'm not sure of that yet. Not convinced. Life is the most amazing thing! I loved physics and thought I would never be interested in anything else, but now I want to write. For example, about the fact that man, hot-blooded man, is not right for science. He's a hindrance to it. Tiny man with his tiny problems. Or about how a few physicists could change the entire world. A new dictatorship. A dictatorship of physics and mathematics. A new career has opened up for me.

Before my operation – I knew I had cancer – I thought I only had days, a very few days, left to live, and I desperately wanted not to die. I was suddenly seeing every leaf, bright colours, a bright sky, the vivid grey of tarmac, the cracks in it with ants clambering about in them. 'No,' I thought to myself, 'I need to walk round them.' I pitied them. I did not want them to die. The aroma of the forest made me feel dizzy. I perceived smell more vividly than colour. Light birch trees, ponderous firs. Was I never to see this any more? I wanted to live a second, a minute longer! Why had I spent so much time, so many hours and days, sitting in front of the television surrounded by piles of newspapers? What matters most is life and death. Nothing else exists. Nothing to throw on the scales.

I have understood that only the time you are living has any meaning. The time of our lives.

> *Valentin Alexeyevich Borisevich, former laboratory director of the*
> *Institute of Atomic Energy, Belarus Academy of Sciences*

Monologue on something more remote than Kolyma, Auschwitz and the Holocaust

There are things I have to say . . . My emotions are overwhelming . . .

During the first days, my feelings were mixed. I remember the two most powerful were fear and resentment. Everything had happened, and no one was telling us anything: the authorities were silent, the doctors were saying nothing. There were no answers. At district level they were awaiting instructions from the provincial level, at that level they were awaiting instructions from Minsk, and in Minsk from Moscow. A long, long chain . . . which meant that in effect we had no protection. That was the main feeling during those days. Somewhere, far away, there was Gorbachev and a few others. Two or three people deciding our fate. The fate of millions of people. And just a few people could kill us. Not maniacs and criminals with a terrorist plan in their heads, just ordinary operators on duty that day at an atomic power station. They were probably quite decent men. When I thought that, it came as such a shock. This was something I discovered for myself. I realized that Chernobyl was more remote than Kolyma, Auschwitz and the Holocaust. Am I making sense? Someone with an axe, or a bow and arrow, or a man with a grenade launcher or gas chamber, could not kill everyone, but a man with the atom . . . That means the whole world is in danger.

I'm not a philosopher so I won't try philosophizing. I'll share what I remember.

The panic during those first days: some rushed to the pharmacy and bought up stocks of iodine. Some people stopped visiting the market to buy milk and meat, especially beef. In our family, we stopped trying to be economical and bought more expensive sausage, hoping that meant it was made from safe meat. We soon found out, though, that they were deliberately adding contaminated meat to expensive sausage. Their logic was that, as it was expensive, people would buy only a little of it and eat less. We had no protection. But of course you must already know all this. I want to tell you about something else. The fact that we were the Soviet generation.

My friends were doctors and teachers, the local intelligentsia. We had our own circle, gathered at my house, drank coffee. Two bosom friends are sitting there, one a doctor. Both have small children.

The first: 'I'm going tomorrow to stay with my parents. I'm getting the children out. If they suddenly became ill I would never forgive myself.'

The second: 'The newspapers are saying the situation will be back to normal in a few days. We've got the army there, helicopters, armoured vehicles. It was announced on the radio that . . .'

The first: 'I would advise you to do the same. Get your children.

Take them away! Hide them! Something more terrible than war has happened. We can't even imagine what.'

They suddenly became shrill and ended up quarrelling, flinging accusations at each other:

'Where's your maternal instinct? You're a mindless fanatic!'

'You're a shameless traitor! Where would we be if everyone behaved like that? Would we have won the war?'

A quarrel between two beautiful young women who passionately loved their children. It all seemed so familiar.

Everybody there, including me, felt she was spreading panic, rocking the boat, undermining our confidence in everything we were accustomed to trust. We should wait to be told. They would announce what was to be done. She was a doctor, though, and knew more than we did. 'You're incapable of protecting even your own children! If there's no threat, why are you so scared!'

How we despised, even hated, her at that moment! She had spoiled our evening. Am I making this clear? It was not only the authorities who were deceiving us: we ourselves didn't want to know the truth, somewhere deep down, at a subconscious level. Now, of course, we don't want to admit that to ourselves. We prefer to curse Gorbachev, to curse the Communists . . . It was all their fault. We were blameless. We were victims.

The next day, she left, and we dressed our children up and took them to the May Day Parade. We didn't have to go. We had a choice. We were not forced, nobody required our attendance. But we considered it was our duty. What? At a time like this, on a day like this? We should all be together. We ran down into the street and joined the throng.

On the podium, there were all the secretaries of district Party committees, standing shoulder to shoulder with the first secretary and his little daughter. She was standing somewhere she could be seen. She was wearing a coat and hat, even though the sun was shining, and he was wearing a military waterproof cape. But there they were. I do remember that. The contamination was not only in our land but in our minds. That had been going on, and would continue, for years.

During that period, I changed more than during all the previous forty years of my life. We are locked in a zone. The resettlement programme has been discontinued and it's as if we are living in a Gulag. The Chernobyl Gulag. I work as a children's librarian. Children want to talk. Chernobyl is everywhere, all around us. We have no choice but to learn to live with it. Especially the children in the senior classes. They have questions. What can they do? How can they find things out? Read? There are no books. Watch movies? There are none. There aren't even fairy tales or myths.

I teach through love. I want to conquer fear through love. There I am, standing in front of the children: I love our village, our little river, our forests; they are just the best! I know nothing better. I am not deceiving them. I teach with love. Am I making this clear?

I have a problem because of my years of teaching. I always speak and write a bit loftily, more emotionally than is fashionable nowadays, but I want to answer your question of why we feel powerless. I do feel powerless . . . There is our pre-Chernobyl culture, but we do not have a post-Chernobyl culture. We live surrounded by thoughts of war, the collapse of Socialism and an uncertain future. There is a lack of new ideas, of new goals and ideas. Where are our writers and philosophers? I will say nothing about the fact that our intelligentsia, who more than anybody longed for and paved the way for freedom, are now elbowed aside, impoverished and humiliated. Nobody has a use for us. We're simply not needed. I can't afford to buy even essential books, and books are my life. I . . . we . . . More than ever we need new books, because all around us there is a new life. We are not part of it, and can't reconcile ourselves to that. I keep asking myself: why? Who is going to do our work? Television does not teach children, that's something you need teachers for. But that is a separate subject.

I've recalled all this in the interests of knowing the truth about those days and about our feelings. So as not to forget how much we have changed. And how much our life has changed . . .

Lyudmila Dmitrievna Polyanskaya, village schoolteacher

Monologue on freedom and the wish to die an ordinary death

It was freedom. I felt myself a free man there . . .

Are you surprised? I can see you are. It's something you can understand only if you were in the war. They get drunk, men who fought in the war, and start reminiscing. I've listened to them and know they still look back nostalgically to those years. To the freedom, to the elation. 'Not one step back!' Stalin's order. The NKVD detachments who shot anyone who tried to retreat. We know the story. It's all history now. But you shoot, survive, get your authorized hundred mils of vodka, your tobacco ration. There are a thousand times you might be killed, blown to pieces, but if you make the effort, if you outwit the Devil, the sergeant, the battalion commander, the man wearing an odd helmet and wielding a different-looking bayonet, if you talk your way out of trouble with the Almighty Himself, you can survive!

I was at the reactor. I was there, like a soldier in a trench right on the front line. Terror and freedom! Living life at full throttle. You can't imagine that in ordinary life, can't feel what it's like. I remember we were constantly being readied for war, but when the call came we were not prepared. I was not ready. That day, I was planning to take my wife to the cinema in the evening. Two army types turned up at the factory and called me out. 'Can you tell the difference between diesel and petrol?' 'Where are you sending me?' I asked. 'Where do you think? To Chernobyl, as a volunteer.' My military profession was as an expert in rocket fuel. A classified speciality. They took me straight from the factory, in a vest and T-shirt, didn't let me look in at home. I asked them to let me warn my wife. 'We'll take care of that.' There were about fifteen of us in the bus, reserve officers. I liked the look of them. Their attitude was, 'If we're called, we go; if it's needed, we do the job.' If they took us to the reactor, we'd go up on the roof.

Watchtowers at the evacuated villages, armed soldiers up there. Assault rifles with live ammunition. Road barriers. Notices: 'Verge contaminated. Entry or stopping strictly prohibited.' Trees a grey-white colour from the decontaminant spray, which was as white as snow. Your brain out of kilter straight away! For the first few days, we were afraid to sit on the ground or the grass, ran everywhere instead of walking, pulled on respirators the minute a vehicle passed by. After the shift, we sat in our tents. Ha ha! Within a couple of months, it all seemed normal, it was just your life. We picked plums, trawled for fish. What amazing pike there were! And bream. We dried the bream to go with our beer. You've probably heard all this before. We played football. Went swimming! Ha ha . . . Believed in fate; deep down we were all fatalists, not chemists. Not rationalists. The Slav mentality. I believed in my lucky star! Ha ha! Now I'm second category disabled. I fell ill immediately. Damned X-rays. Clear enough. I didn't even have a medical record at the clinic before that. Damn it! I'm not the only one. It's the mentality . . .

I'm a soldier. I sealed other people's houses, went into other people's homes. It's a peculiar feeling, like you're prying . . . Or there's the land no one can sow on, a cow butting at a closed gate. The house is padlocked and the cow's milk is dripping on the ground. A peculiar feeling. In the villages not yet evacuated, peasants were busy distilling moonshine. That was their livelihood now. They sold it to us, and we had so much money: triple time at work, and triple living expenses. Then there was an order that anyone who was drinking would be kept on for a second term. So was vodka good for you or not? Psychologically at least? The men there certainly believed it was. It seemed obvious.

The peasant's life was very straightforward: you sowed something, it grew, you harvested it, and everything else was someone else's concern. They had no interest whatever in tsars or regimes, in whether they were being ruled by a general secretary of the Central Committee or a president, or in spacecraft or atomic power stations, or protest rallies in the capital. They couldn't believe their world had been turned upside down in a single day and that now they were living in a different world: the world of Chernobyl. They had no intention of going anywhere else. Some of them became ill just from the shock. They didn't resign themselves to it, they wanted to live as they always had. They secretly brought firewood back; they picked green tomatoes and pickled them. If the jars exploded, they boiled them again. How could you destroy that, bury it, turn it into garbage? But that was precisely what we were doing. Nullifying their work and the time-honoured meaning of their lives. For them, we were the enemy.

I couldn't wait to get to the reactor itself. 'No need to hurry,' I was warned. 'In the last month before demob, they have everyone up there on the roof.' We were serving for six months. Sure enough, after five months we were redeployed, right up to the reactor. There were various jokes, and serious talk that we would have to pass the test of walking across the roof. We thought we might last five years after that. Seven? Ten? Fair enough. For some reason, mostly they reckoned it was five. Where did that come from? There was no fuss, no panic. 'Volunteers, one step forward!' The whole company took a step forward. There was a television monitor in front of the commander. He turned it on. The screen showed the roof of the reactor: lumps of graphite, melted bitumen. 'Right, lads, you can see this debris lying around. Clear it away. And here, in this square, you need to break a hole through.' The permissible time to be exposed was forty or fifty seconds. That was according to the instruction book, but it was impossible. It required at least several minutes. You had to get there and back, make a run and tip the stuff down. One loaded the litter, the others emptied it out, down there, into the ruins, into the hole. You had to tip it in but not look down. People did, of course.

They wrote in the newspapers, 'The air above the reactor is clean.' We read that and laughed, and swore with gusto. The air might be clean, but what kind of dose were we being exposed to? They issued radiation meters. One was calibrated up to five roentgens. That was off the scale in the first minute. Another was like a fountain pen and went up to a hundred roentgens. In certain places, even that one went off the scale. For five years, they said, you won't be able to have children. That seemed to assume we wouldn't die within five years. Ha ha! . . . There were various jokes, but no fuss, no panic. Five years. I've already lived for ten . . . Ha ha! . . . They presented us with certificates. I have two . . . With all those pictures of Marx, Engels and Lenin. Red flags.

One lad disappeared. We thought he'd done a runner. Two days later, he was found in the bushes. He'd hanged himself. We were all feeling down, you can imagine. Then the deputy political officer gave a speech and claimed the boy had had a letter from home saying his wife had been unfaithful. Who knows? We were due to be demobbed a week later, and they found him in the bushes. We had a cook. He was so scared he didn't live in a tent but in the stores. He dug himself a niche under boxes of butter and tinned stew, moved his mattress in there, a pillow, and lived underground. An instruction arrived to put together a new team to go up on the roof. We had all been up there already. Just find people! So they put him down on the list. He only went up there once, but now he's second category disabled too. Often phones me. We keep in touch, support each other, to keep the memory alive. It will last for just as long as we stay alive. Write that down.

What the newspapers print is a pack of lies. Lies from start to finish. I've never read anywhere about how we made ourselves chain mail, lead shirts and pants. We were issued rubber gowns impregnated with lead, but we fashioned lead trunks for ourselves. We tried to protect ourselves. Of course we did. In one village, we were shown two secret brothels. Everybody went there. Men taken away from home, six months without a woman . . . Emergency! The local girls were up for it too. 'We're going to die soon anyway,' they said, crying. You wore your lead trunks over your underpants. Write that. We told all these jokes. Here's one. They send an American robot up to work on the roof. It operates for five minutes, then breaks down. Then a Japanese robot lasts nine minutes before it breaks down too. The Russian robot works for two hours, then, over the walkie-talkie, 'Okay, Private Ivanov, you can come down now for a cigarette break.' Ha ha!

Before we were to go up on to the reactor, the commander was giving us our orders. We were lined up. Several of the lads rebelled:

'We've already been up there, you're supposed to send us home now.' My job was supposed to be looking after the fuel, petrol, but I got sent to the roof too. I didn't complain. I wanted to go there myself, to take a look. Anyway, they rebelled. The commander says, 'We will be sending volunteers on to the roof. Others step out of the line. The prosecutor will be having a word with you.' Well, these lads stood there, talked among themselves, and consented. If you've taken the oath, kissed the flag, gone down on your knee before the banner . . . I don't think any of us were in any doubt they were well capable of court-martialling us and putting us in prison. They put the word round that you would get two or three years. If a soldier was subjected to more than twenty-five roentgens, the commander could be court-martialled for exposing his men to excessive radiation. Of course, nobody ever went over twenty-five roentgens. Everybody had less. You get my drift? But I really liked the lads. Two were sick, but one of them said, 'I'll go.' He'd already been up on the roof once that day. They recognized that. He got a bonus: 500 roubles. Another was hammering away up there. It was time to go, but he carried right on hammering. We waved to him to come off, but he got down on his knees and finished the job. The roof had to be broken through at that point to insert a chute for the rubbish to go down. He wouldn't get up until he had broken through. He got a thousand-rouble bonus. For that money, you could buy two motorbikes at the time. He's now a category one invalid. No arguing with that. They paid you for fear immediately. Now he's dying. He's in terrible pain. I went to see him this weekend. 'Ask me what I really want?' 'What?' 'An ordinary death.' He's forty. Loved women. Has a beautiful wife.

Demob. We all piled into the vehicles. For as long as we were driving through the exclusion zone, we kept honking the horns. When I look back on those days . . . I was there, next to something big, something unimaginable. Words like 'gigantic', 'unimaginable' don't convey it. It felt like – what? (*Ponders.*)

I've never experienced anything like it, even in love.

Alexander Kudryagin, clean-up worker

Monologue on a freak who is going to be loved anyway

Don't be embarrassed. Ask away. We've already had so much written about us, we're quite used to it. Sometimes they remember to send us the newspaper, but I don't read it. Who's going to understand? You need to live here . . .

My daughter said recently, 'Mum, if I give birth to a freak, I'll love it anyway.' Can you imagine? She's in tenth grade at school and already thinking things like that. Her friends, they're all thinking the same way. Some friends of ours had a baby boy. They were so looking forward to him, their first child. A lovely young couple. But their boy has a mouth from ear to ear, or would have, only one ear is entirely missing. I don't visit them as much as I used to. I can't bear it. But my daughter goes to see them occasionally. She's drawn there. I don't know whether she's trying to get used to the idea, or trying to see whether she'll measure up to the challenge. But I can't cope.

We had the choice of moving away, but my husband and I thought it over and turned it down. We're afraid of other people; whereas here, we're all just the people of Chernobyl, together. We're not afraid of each other. If someone offers you apples or cucumbers from their plot, you accept them and eat them. We don't politely put them away in a pocket or bag and throw them away afterwards. We have a shared memory, the same fate. And anywhere else we're regarded as outsiders. People look askance at us, fearfully. Everybody is so used to the words 'Chernobyl', 'Chernobyl children', 'Chernobyl evacuees'. 'Chernobyl': now that gets prefixed to everything about us. But you don't know the first thing about us. You're afraid of us. You run away. If we weren't allowed out of here, if they put a police cordon round us, many of you would probably be relieved. (*Stops.*)

Don't try to change my mind about anything, don't try to talk me round! I learned and experienced all that at the very beginning. I seized my daughter and fled to Minsk, to my sister. My own sister wouldn't let us in the door because she had a baby she was breast-feeding. I could never have imagined that in my worst nightmare! I couldn't have made it up. We spent the night at the railway station.

All sorts of crazy ideas came into my mind. Where could we go? Perhaps it would be better to end it all, rather than live a life of suffering. Those were the first days. Everybody was picturing some sort of dreadful illness. Unimaginable sicknesses. And I'm a doctor. You can only try to guess at what must have been passing through other people's minds. Rumours are always worse than accurate information, however little. I look at our children: wherever they go, they feel like outcasts. Something out of a horror story, a target for jibes. In a Young Pioneers' camp where my daughter went on holiday one year, the other children were afraid to touch her: 'Chernobyl Glow-worm!' 'She glows in the dark.' They called her outside one evening to check. They wanted to see if she had a halo over her head.

They say it's like the war. The war generation. That's who they compare us with. But they were lucky. For them, it ended with victory. They won! That fired them up, to use today's way of speaking. A tremendously strong determination to survive. They had nothing to fear. They wanted to live, and study, and have babies. But what about us? We're afraid of everything, afraid for our children, afraid for our unborn grandchildren. They haven't even been born, but already we're frightened. People smile less now, don't sing the way they used to at holiday time. It's not only the landscape that changes when forests and scrub spring up where fields used to be: the national character changes. Everybody suffers from depression, a sense of doom. For some people, Chernobyl is a metaphor, or a slogan. But that is where we live. It's just where we live.

At times, I think it would be better if you didn't write about us at all. Didn't view us from the sidelines, didn't try to diagnose us with radiophobia or whatever, didn't separate us out from everybody else. Then people wouldn't be so afraid of us. After all, you don't talk about a cancer patient's dreadful disease in his own home. And you don't mention someone's sentence in their cell, when they're in prison for life. (*Silence.*)

I've talked so much. I don't know whether you need that or not. Shall I lay the table? Shall we have lunch, or are you afraid? Tell me straight out. We don't get offended any more. We've seen it all! A reporter came to interview me. I could see he was thirsty. I brought him a mug of water, but he took his own water out of a bag. Mineral water. He was embarrassed, started making excuses. Needless to say, our interview went nowhere. I couldn't be open with him. I'm not a robot or a computer, or a lump of metal! He thought he could sit there drinking his mineral water, being afraid even to touch my mug, and I was supposed to pour out my heart to him, let him into my soul.

(*Now we are sitting at the table, having lunch.*) I cried all night yesterday. My husband recalled, 'You used to be so pretty.' I know what he was on about. I see myself in the mirror. Every morning. People age prematurely here. I'm forty, and you would think I was sixty. That's why the young girls are in such a hurry to get married. They're sad about their youth being so brief. (*Breaks down.*) Well, what do you know about Chernobyl? What can you write down? Forgive me. (*Silence.*)

How can you write down my soul? I can't always make sense of it myself . . .

Nadezhda Afanasyevna Burakova, resident of Khoyniki

Monologue on the need to add something to everyday life in order to understand it

Do you need facts, details of those days? Or my story? It was there I became a photographer . . .

Until then, I'd never had any interest in photography, but there I suddenly began taking pictures. I happened to have a camera with me. Thought I would just take some photos for my own satisfaction, but now it's my profession. I couldn't rid myself of the new feelings I experienced. These were not fleeting experiences, but the whole story of my soul. I changed . . . I saw a different world. Know what I mean?

(*While he is speaking, he is laying photographs out on the table, on chairs, on the windowsill: a gigantic sunflower the size of a cartwheel; a stork's nest in a deserted village; a lonely village graveyard with a sign on* the gate, 'High radiation levels. Entry by vehicle or on foot prohibited'; a pram in the yard of a house with boarded-up windows, with a crow sitting on it as if this was its nest; an aeons-old formation of cranes flying over neglected fields.)

People ask why I don't use colour film, but Chernobyl in Russian means 'a black story'. The other colours don't exist there. My story? It's a commentary to go with this . . . (*Points to the photos.*) Okay, I'll try. Although, you know, it's all here . . . (*Again points to the photographs.*) At the time, I was working in a factory, and studying history externally at university. I was a metal-worker, second class. We were formed into a group and sent off urgently. As if we were going to the front.

'Where are we going?'

'Where you're ordered to go.'

'What are we going to be doing.'

'What you're ordered to do.'

'But we're construction workers.'

'And that is what you'll be doing. Constructing.'

We built outbuildings: laundries, storehouses, sheds. I was detailed to unload cement. What cement, or where it came from, no one checked. We loaded it and unloaded it. I spent the day shovelling it, and by evening only my teeth were white. A cement man. Grey. My body and my boiler suit, all through them. In the evening, I shook them out, you know, and in the morning I put them back on. We were treated to indoctrination talks. Heroes, feats of heroism, on the front line. All that military language. But as for what a rem, or a curie, or a milliroentgen might be? We asked the commander, but he couldn't explain. They hadn't taught him that at his military college. Milli-, micro-, it was all Chinese to him. 'What do you want to know that for? Carry out orders. You're soldiers here.' Yes, soldiers, but not *zeks* in a prison camp.

A commission arrived to inspect us. 'Don't worry,' they assured us, 'everything is fine here. Background radiation is normal. Just four kilometres from here is not habitable, the people will be evacuated. But here there's nothing to worry about.' They had a radiation monitoring technician with them. He went and switched on a box slung over his shoulder and ran a long rod over our boots. He jumped back. An involuntary reflex.

This is where it starts getting interesting, especially for you as a writer. How long do you think we remembered that incident? Barely a few days. After all, we Russians are incapable of thinking only about ourselves and our own lives, of not looking beyond that sort of thing. Our politicians are incapable of thinking about the value of a life, but that goes for us as individuals too. Know what I mean? We just don't think that way. We're made of different stuff. Of course, we all got drunk there, seriously drunk. By night, no one was sober; we drank, though not to get drunk, but to get talking. After the first couple of shots, someone would get sentimental and start remembering his wife and children, talk about his job and curse those in charge of us. But then, after a bottle or two, all the talk was about the country's future and the nature of the universe. We argued about the merits of Gorbachev against Ligachov. About Stalin. About whether or not we were a great power, and whether we would overtake the Americans or not. It was 1986. Whose planes were better, whose spacecraft were more reliable? Okay, so Chernobyl had blown up, but our Gagarin was the first man to fly in space! Know what I mean? Until we were hoarse, until it was morning. Why we hadn't been issued radiation meters or anti-radiation pills just in case, or why there were no washing machines to clean our boiler suits every day rather than twice a month – those were the least of our concerns. Just mentioned in passing. Hell, that's the kind of people we are!

Vodka was worth more than gold. It was impossible to buy. We'd drunk everything we could get in the villages round about: vodka, moonshine, aftershave. We even went as far as varnishes and aerosols. There would be a three-litre jar of moonshine on the table, or a string bag full of bottles of chypre cologne. And there was talk, talk without end. Our team included teachers and engineers, a complete Communist International: Russians and Belarusians, Kazakhs and Ukrainians. There were philosophical conversations about how we were enslaved by materialism, which confined us to the material world, while Chernobyl was a path to infinity.

I remember us debating the course of Russian culture with its penchant for tragedy. Without the overhanging shadow of death, nothing would be understood. It would be possible to come to terms with the disaster only by building on the foundation provided by Russian culture. Only it was up to the job, had been full of premonitions of this. We had been afraid of the atomic bomb and mushroom clouds, but what had happened? . . . Hiroshima had been terrifying, but at least it was comprehensible. But this . . . We know how a house can be set alight by a match or an exploding shell, but this was like nothing we knew. We heard rumours that the fire was unearthly, not even fire but light. A glimmering. A radiance. Not blue, but a translucent azure. And without smoke. If scientists had been sitting on the throne of the gods, now they were fallen angels, demons! Human nature had remained as much of a mystery to them as ever. I am Russian, from Bryansk. We had an old man there, sitting outside his door. The house was leaning to one side and about to collapse, but there he sat, philosophizing, putting the world to rights. Know what I mean? You can guarantee every factory smoking room will have its Aristotle. Every bar. And there we were, a stone's throw away from a damaged atomic reactor . . .

Newspaper reporters flocked to us, took photographs with cheap effects. The window of an abandoned home: they would put a violin on the sill and call it *Chernobyl Symphony*. There was no need to invent anything. I wanted everything to be remembered: the globe of the earth in a school yard, crushed by a tractor; blackened washing which had been hanging for several years on a balcony to dry; dolls which had grown old in the rain. Neglected mass graves from the war, the grass on them as tall as the plaster soldiers, birds nesting on their plaster rifles. A door smashed in, the house ransacked by looters, with the curtains drawn across its windows. People had gone, leaving only their photographs living on in their homes, as if they were their souls. Nothing was insignificant or trivial. Everything needed to be remembered, accurately, in detail: the time of day when I saw it, the colour of the sky, the sensations. Know what I mean? Someone had left this place behind forever. What did that mean? We are the first people to have experienced that 'forever'. We

can't allow ourselves to miss a single detail. The faces of old peasants like the faces in icons . . . They, least of all, knew the meaning
of what had happened. They had never left their farm, their land.
They came into the world, loved, ate bread earned by the sweat of
their brow, and procreated. They hoped to live to see grandchildren
and, when life was over, meekly departed this earth, returned to the
soil, became part of it. The Belarusian cottage! To us town-dwellers,
a house is a machine for living in. For them, though, it's their whole
world. A cosmos. You travel through the empty villages and so want
to meet a human being. A plundered church . . . We went in, and it
smelled of wax. You wanted to pray . . .

I needed to remember all this. I started taking pictures. That's
my story.

I went recently to the funeral of a friend I had been there with.
He died of leukaemia. The wake. Following Slavic custom, we
drank and ate, know what I mean? And talked until midnight. At
first, about him, the departed. But then? Once again about Russia's
destiny and the state of the universe. Would Russian troops be
withdrawn from Chechnya? Would there be a second Caucasian
War? Had it begun already? What were the chances of far-right
Zhirinovsky becoming president? What were the odds on Yeltsin?
We talked about the British monarchy and Princess Diana. About
Russian monarchy. About Chernobyl. That led us off into speculation. One theory was that extraterrestrials knew about the disaster
and were helping us out; another was that this had been a cosmic
experiment, and that eventually genius children with extraordi-
nary abilities would start to be born. Alternatively, that the
Belarusians would disappear, as other peoples had: the Scythians,
Khazars, Sarmatians, Cimmerians and Huastecs.

We are adepts of metaphysics. We live not on the ground but in
the realm of dreams, of talk, of words. We need to add something
to everyday life in order to understand it. Even when we are living
next to death. That's my story. Now I have told it. Why did I take
up photography? Because words were not enough . . .

Viktor Latun, photographer

Monologue on a mute soldier

I won't be going back into the Zone itself, although in the past I was drawn to it. If I see it and think about it, I will become ill and die. My imaginings will die with me.

Do you remember a war film called *Come and See*? It came out in 1985, and I couldn't watch it through to the end. I fainted. They killed a cow in it. The pupil of its eye filled the screen. Just the pupil. I didn't last to see people being killed. No! Art is love. I'm totally convinced of that! I'm reluctant to turn on the television or read the newspapers today. It's all about killing, killing, killing. In Chechnya, in Bosnia, in Afghanistan. I'm losing my mind. My eyesight is deteriorating. Horror has become commonplace, even banal. And how we've changed, so that today's horror on the screen has to be more dreadful than yesterday's. Otherwise we don't find it frightening. We've gone too far.

Yesterday, I was on a trolleybus. There was a little incident: a boy wouldn't give up his seat to an old man, who was telling him off: 'When you're old, you will find nobody stands up for you.'

'I'm never going to be old,' the boy retorts.

'How so?'

'We're all going to die soon.'

All around you, people are talking about death. Children are thinking about it. But that's something you should contemplate at the end of life, not when it's just beginning.

I see the world as made up of incidents. For me, the street is theatre, home is theatre. Man is theatre. I never remember an event in its entirety. Only details, gestures . . . Everything is muddled, mixed up in my mind. I'm not sure if I saw something in a film or read it in the newspapers. Or whether I saw or heard it somewhere myself, glimpsed it. I see a mad fox walking down an abandoned village street. It's gentle and kind, like a child. It rubs up against feral cats, and chickens.

Silence. There's such a stillness there! Completely different from what we have here. And then suddenly it's broken by a strange human voice: 'Gosha's a pretty boy, Gosha's a pretty boy.' A rusty cage with the door open is swinging on an old apple tree. A pet parrot is talking to itself.

The evacuation is beginning. The school has been closed and sealed, the collective-farm office, the village soviet. During the day, soldiers took away safes and documents, and at night the villagers come to help themselves to whatever has been left. They pilfer books from the library, mirrors, chairs, sanitary fittings, a heavy globe. Some latecomer rushes in next morning, but everything has already gone. He purloins empty test tubes from the chemistry laboratory. Although they all know they're being evacuated themselves in three days' time and that they will have to leave everything behind.

Why am I gathering, collecting all this? I'm never going to direct a play about Chernobyl, just as I've never staged a single play about war. I'm never going to show a dead person on the stage. Not even a dead bird or animal. In the forest, I came upon a pine tree. There was something white. I thought it was mushrooms, but it was dead sparrows with their little breasts facing upwards. There, in the Zone . . . I don't understand death. I don't look at it, in order not to go crazy. So as not to cross over, into that other side of life. War should be depicted so horribly it makes you sick, so it makes you ill. It's not something to be watched . . .

In those first days . . . They hadn't yet let us see a single picture, but I was already imagining: collapsed roofs, shattered walls, smoke, broken windows. Hushed children being taken away. Queues of vehicles. The adults crying, but not the children. They hadn't yet printed a single photograph . . . Probably, if you were to ask people, that's the only image of terror we have: an explosion, a fire, dead bodies, panic. I remember all that from my childhood. (*Trails off.*) We can talk about that later, on another occasion. But here, what had happened was something we didn't know about. A different kind of fear. This was something you couldn't hear or see. It had no smell, no colour, and it changed us physically and mentally. The composition of our blood changes, our genetic code changes, the natural scenery around us changes. And it makes no difference what we think or do. Here we are: I get up in the morning, have a mug of tea and go to a rehearsal with my students. And all the time, I have this hanging over me. Like a portent. Like a question. I have nothing to compare it to. What I remember from my childhood is quite different.

I've seen only one good film about the war. I can't remember the title. It was about a mute soldier. He didn't speak a word throughout the film. He was transporting a pregnant German woman, pregnant by a Russian soldier. The baby was born, born on their journey, in their cart. He lifts it up in his arms and holds it, and the baby pees on his rifle. The man laughs. It's as if that's the way he speaks, through that laughter. He looks at the baby, he looks at the rifle, and he laughs. The end.

In that film, no one's a Russian, no one's a German. There's a monster: war; and there's a miracle: life. Now, though, after Chernobyl, everything has changed. The world has changed too. It no longer seems eternal. The earth seems to have shrunk: it's small now. Our immortality has been taken away from us. That is what has happened. We've lost our sense of eternity. Meanwhile, what I see on television is people killing, every day. Shooting. One person kills another. After the Chernobyl disaster.

Something very blurred, as if seen from a great distance . . . I was three when I and my mother were taken to Germany. To a concentration camp. I remember everything that was pretty. Perhaps that's just the way I see. A high mountain. Rain falling, or perhaps it was snow. People standing in an enormous black semicircle, all of them with numbers. Numbers on their shoes. So precise, such bright yellow paint on their shoes. And on their backs. Numbers everywhere, numbers. And barbed wire. A man wearing a helmet up in a watchtower. Dogs running about, barking very, very loudly. But there is no fear. Two Germans. One is a big fat man dressed in black, the other one is small – in a brown costume. The one in black indicates somewhere with his hand. A black shadow comes out from the dark semicircle and turns into a person. The German in black starts hitting him. Rain is falling, or perhaps it's snow. Falling . . .

I remember a tall, handsome Italian man. He was singing all the time. It made my mum cry. Other people too. I couldn't understand why everyone was crying when he sang so beautifully.

I wrote some short pieces about the war. Tried to. They didn't amount to anything. I'll never stage a play about war. It just wouldn't work.

We took a cheery play called *Give Us Water, Well* into the Chernobyl Zone. A folk tale. We arrived at Khotimsk, the district centre. They have a children's home there for orphans. Nobody thought to evacuate them.

End of part one. They didn't clap. They didn't stand up. Silence. Part two. End of the play. Again, no clapping, and they didn't stand up. Silence.

My students were in tears. They gathered behind the scenes. What was wrong? Then it dawned on us: they believed everything that was happening on the stage was real. All through the play, they were waiting for a miracle to happen. Ordinary children, well brought up children, would have known they were only watching a play. But these children were waiting for a miracle.

We Belarusians have never had anything permanent. Even land was not permanent: it was constantly being seized by somebody who erased all trace of us. It meant that we ourselves had no permanence to live in, the way it's written in the Old Testament that this one begat that one, and that one begat someone else. A chain, links. We don't know how to handle permanence. We can't live with it, can't make sense of it. But now, finally, it has been bestowed upon us. Our permanence is – Chernobyl. Here we have it. And what do we do? We laugh about it, like in the old parable. People are commiserating with someone whose house and barn have burned down. Everything has gone up in flames, but he answers, 'Then again, think how many mice it exterminated!' and cheerily swipes the floor with his hat. That's the archetypal Belarusian for you! Laughter through tears.

But our gods don't laugh. They are martyrs. It was the ancient Greeks had gods who laughed and made merry. But what if fantasies, dreams and jokes are also texts, about who we are? Only we're no good at reading them? There's one melody I hear, always and everywhere. It goes on and on. It's not really a tune or a song, but a keening. It is a pre-programmed aptitude of our people to attract misfortune, a never-failing expectation of woe. Happiness? Happiness is temporary, accidental. We have folk sayings: 'One disaster doesn't count', 'A stick is no defence against disaster', 'You'll lose the race from a punch in the face', 'Forget Christmas cheer if disaster is here'. Other than suffering, we have nothing. No other history, no other culture.

But still my students fall in love and have babies. Only their babies are quiet and puny. After the war, I came back from that concentration camp . . . I'm alive! All we needed at that time was to survive. My generation is still surprised that it did. If there was no water, I could eat snow. In the summer, I was never out of the river: I would dive in a hundred times. Their children will never be able to eat snow. Even the purest-looking, whitest snow . . . (*She is lost in thought.*)

How do I imagine the play? I am thinking about it, you know, all the time.

From the Zone they brought me one plotline. A modern folk tale.

An old man and an old woman stayed behind in their village. In winter, the old man dies. The old lady is the only person at his funeral. For a week, she hacks out a pit among the graves. She wraps her husband in a warm shroud against the frost, lays him on a children's sledge and pulls him along. All the way, she talks to him, recalling their life together. She roasts their last chicken for the wake. A hungry puppy homes in on the smell and joins her, so she has someone to talk to and cry with.

I once even dreamed about my future play.

I saw an empty village, the apple trees in blossom, the cherries flowering. So luxuriant, so bright and cheerful. The wild pear blooming in the graveyard . . . Cats are running through the overgrown streets with their tails held high. There is nobody here. They mate. Everything is flowering. Such beauty and stillness. Then the cats run into the street, expecting someone. They probably still remember human beings . . .

We Belarusians have no Tolstoy, no Pushkin. What we have

is Yanka Kupala, Yakub Kolas. They wrote about the land. We are people of the earth, not heaven. Our monoculture is the potato. We dig it, plant it, and all the time look down towards the ground. If anyone does throw back their head, it is to look no higher than a stork's nest. As far as they are concerned, that is high enough. For them, that counts as the sky. A sky called the 'cosmos' is something we don't have, it's absent from our thinking. When we need that, we get it from Russian literature, or Polish. Just as the Norwegians needed Grieg, and the Jews Sholem Aleichem, as a crystal around which they could grow and become conscious of themselves, now we have – Chernobyl! It's sculpting something out of us. Creating. Now we have become a people. The people of Chernobyl. Not just a stretch of the road from Russia to Europe or from Europe to Russia. Only now . . .

Art is remembrance. Remembrance of our having existed. I am afraid, afraid of one thing: that in our lives fear is replacing love . . .

Lilia Mikhailovna Kuzmenkova, lecturer at
Mogilyov College of Culture and Enlightenment, theatre director

Monologue on the eternal, accursed questions: 'What is to be done?' and 'Who is to blame?'

I'm a man of my time. A committed Communist . . .

Nobody lets us speak. That's today's fashion. Today, it's fashionable to blame everything on the Communists. Now we're the Enemies of the People, criminals all of us. Now we carry the can for everything, even the laws of physics. At that time, I was first secretary of the District Party Committee. The newspapers write now that the Communists are to blame for building inferior, cheap atomic power stations with no concern for human life. They ignored human beings. For them, people were just sand, the manure of history. To hell with them! Damn them! It's those accursed questions: 'What is to be done?' and 'Who is to blame?' Eternal, unchanging throughout our history. People can hardly wait. They're thirsting for revenge and blood. To hell with the

Communists! Damn them! These people are just lusting for severed heads . . . Bread and circuses . . .

Others are saying nothing, but I'll speak out. You write – well, not you specifically, but the newspapers – that the Communists lied to the people; they hid the truth. But we had to. There were telegrams from the Central Committee, from the Provincial Party Committee. We were given the task of avoiding panic. Panic really is a terrible thing. It was only during the war that people were following reports from the front as closely as they were the bulletins from Chernobyl. Fear and rumours. People were killed not by radiation but by an incident. We had to . . . our duty . . .

It's not true to say we were hiding everything from the outset. At first, no one recognized the scale of what was happening. Priority was given to higher political considerations. But if you disregard emotions, disregard politics . . . We have to admit that no one believed what had happened. The scientists themselves couldn't believe it! It was completely unprecedented. Not only in the USSR, but anywhere in the world. The experts at the power station studied the situation, and took decisions on the spot. I recently watched a programme, *Moment of Truth*, which featured Alexander Yakovlev, a member of the Politburo, the Party's chief ideologist at the time, the man next to Gorbachev. What does he remember? Up there, on the summit, they themselves had no understanding of the whole picture. At a meeting of the Politburo, one of the generals said to them: 'Who cares about radiation? On the test site after an atomic explosion . . . that evening we all drank a bottle of red wine and that sorted it.' They were talking about Chernobyl as if it was just another accident, nothing out of the ordinary.

If I'd said then it was wrong to bring the people out in the streets? 'You want to sabotage the May Day Parade? That's a political offence! Put your Party card on the table immediately!' (*Calms down a bit.*) I'm not making that up, you know. It's the truth. The fact of the matter. They say that Shcherbina, the chairman of the government commission, when he arrived at the station in the first days after the explosion, demanded to be taken to the scene immediately. They tried to explain to him: piles of graphite, ridiculous

fields of radiation, high temperatures. You couldn't go near it. 'I
don't care about your physics. I have to see everything with my
own eyes!' he bawled at his subordinates. 'I have a report to deliver
this evening to the Politburo!' Army behaviour patterns. That was
all they knew. They didn't understand that physics really exists. A
chain reaction ... and no amount of orders and government
decrees would alter the laws of physics. That really is what explains
the world, not the ideas of Karl Marx.

But if I'd said then ... If I'd tried to cancel the May Day Parade ...
(*Becoming agitated again.*) The newspapers try to pretend we brought
the people out into the streets while we were sitting safely in underground bunkers. I stood for two hours on that podium, in that sun,
without a hat, without a coat. And on 9 May, Victory Day, I marched
with the ex-servicemen and women. An accordion playing, people
dancing, drinking. We were all part of that system. We believed in
it. We believed in heroic ideals, in the Soviet victory. And that we
would vanquish Chernobyl! We would all pile in and overwhelm it.
We read avidly about the heroic struggle to tame the reactor, which
had broken free of the power of humans. We conducted indoctrination sessions. What are we without ideas? Without some grand
vision? That is no less frightening: look at what is going on now!
Disintegration. Anarchy. Wild West capitalism ... But, the past
has been condemned ... our entire lives. All they talk about is
Stalin, the Gulag archipelago ... but what films we had then! The
joyful songs! Tell me why that was? Answer me ... Think about it,
and give me an answer! Why are there no films like that any more?
No songs? People have to be motivated, inspired. They need ideals.
Then you'll have a strong state. Sausages cannot be an ideal, or a
full fridge. A Mercedes is not an ideal. You need shining ideals! And
that was what we had.

The newspapers ... on the radio and television, they were yelling, 'Give us the truth, the truth!' At demonstrations what was
demanded was: the truth! 'The situation is bad, very bad, very, very
bad! We're all going to die! The nation will cease to exist!' Who
needs that sort of truth? When that mob broke into the Conven-
tion and demanded Robespierre's execution, were they right?

Submitting to the mob, becoming a mob . . . We had to prevent panic . . . My job . . . My duty . . . (*He is silent.*) If I'm a criminal, then why is my granddaughter . . . my child . . . She is ill too . . . My daughter had her in the spring, brought her to us in Slavgorod in nappies. In a pram. They came a few weeks after the explosion at the power station. Helicopters flying, army vehicles on the roads . . . My wife begged me: 'We must send them to our relatives, get them away from here.' I was the first secretary of the District Party Committee. I totally forbade it. 'What will people think if I send my daughter and her baby elsewhere? When their own children have to stay here.' Those who bolted for it, trying to save their own skin . . . I called them into the District Committee, to the office, 'Are you a Communist or not?' It was a test of integrity. If I'm a criminal, why did I not save my own child? (*Becoming incoherent.*) I did . . . she . . . in my own house . . . (*Calms down after a while.*)

Those first months . . . In the Ukraine there was panic, but here in Belarus everything was calm. The sowing season was in full swing. I didn't hide, didn't safely sit it out in an office. I was charging around the fields and meadows. We were ploughing, sowing. You've forgotten that, before Chernobyl, we called the atom 'a peaceful labourer'. We were proud of living in the Atomic Age. I don't recall anyone being fearful. At that time, we didn't yet fear for the future . . . After all, what is a Communist Party first secretary? An ordinary person with an ordinary college degree, most often an engineer or an agronomist. Some had graduated from the Higher Party School. What I knew about radiation was what they'd managed to teach us on civil defence courses. I never heard a word there about caesium in milk, or strontium. We transported milk with caesium to the dairies, delivered meat, scythed grass contaminated with radiation at forty curies. Fulfilled the Plan absolutely conscientiously. I was making sure of that. Nobody had released us from our obligations under the Plan.

Here's another telling detail. In those first days, people were experiencing not only fear, but also elation. I'm someone with absolutely no survival instinct. That's only what you would expect, because I have a very strong sense of duty. There were a lot of us like that then. I certainly wasn't the only one. I had dozens of letters from would-be volunteers on my desk, asking to be sent to Chernobyl. 'At the bidding of their hearts!' People were prepared to sacrifice themselves without a second thought, and without asking anything in return. You can write whatever you like, but the fact of the matter is, there was such a thing as Soviet character. There was such a thing as a Soviet person. It doesn't matter what you write, or how emphatically you deny it . . . You will feel the lack of that kind of person yet, and think back to him nostalgically.

Experts were sent to us, and they got very heated, argued themselves hoarse. I asked one, 'Are our children making sand pies with radioactive sand?' And he came back at me: 'Panic-mongers! Dilettantes! What do you know about radiation? I'm a nuclear expert. We carried out a nuclear explosion. An hour later, I went to the epicentre in a jeep. Over vitrified soil. Why are you spreading panic?' I believed them. I called people to my office and said, 'Comrades, if I run away, if you run away, what will people think of us? They will say the Communists deserted them!' If I couldn't persuade them with words and emotion, I took a different approach: 'Are you a patriot or not? If not, put your Party card on the table. Throw it down!' Some did.

I began to suspect something was wrong. I couldn't help noticing . . . We signed an agreement with the Institute of Atomic Physics to survey our land. They would take samples of grass, layers of black earth back with them to Minsk, analyse it. Then they phoned me: 'Please get some transport organized to take your soil back.' 'Are you joking? It's 400 kilometres to Minsk. I practically dropped the receiver. You want us to cart the soil back?' 'No,' they said, 'we're not joking. Our regulations require these samples to be buried in a safe place, in a reinforced concrete underground bunker. We're having samples brought to us from all over Byelorussia. In a month, we've completely filled up the available capacity.' Do you hear what I'm saying? Here we were, ploughing and sowing that soil. Our children were playing on that land. We were required to deliver milk and meat in accordance with the Plan. People were distilling alcohol from the grain. Apples, pears and cherries were being used for juice.

Evacuation ... If anybody had looked down from above, they would have thought the Third World War had begun. One village was being moved out and another was warned they would be evacuated within a week. And for the whole of that week, they carried on stacking the hay, scything the grass, digging their vegetable plots, chopping firewood ... Life went on as usual. People had no idea what was happening, and a week later they were taken away in army vehicles. Meetings, travelling, indoctrination sessions, sleepless nights. There was just so much going on. I remember a chap stood outside the Municipal Party Committee in Minsk with a placard which read, 'Give the people iodine'. It was hot, but he was wearing a raincoat.

(*We return to the beginning of our conversation.*) You forget ... Back then, atomic power stations were the future. I gave talks, spread the word. I visited one. It was very quiet and imposing, clean. Red flags and pennants. In the corner: 'Victor in Socialist Competition'. Our future. We lived in a fortunate country. We were told we were happy, and we were. I was a free man. I couldn't imagine why anyone would consider my freedom was unfreedom. Now we've been written off by history, as if we don't exist. I'm reading Solzhenitsyn now ... I think ... (*Silence.*) My granddaughter has leukaemia ... I've paid for everything. A high price ...

I'm a man of my time, not a criminal ...

Vladimir Matveyevich Ivanov, former first secretary,
Slavgorod District Party Committee

Monologue of a defender of Soviet power

Hey, fuck off! Hey! (*Air turns blue with swearing.*) Stalin's what you need! The iron fist ...

What are you recording anyway? Who gave you permission? Photographing ... Take that box of tricks of yours away. Or I'll smash it. That's enough, understand? We're living here. Suffering, and you think you'll just write stuff. Hacks! Destabilizing people.

Causing rebellions. That's enough, I said, understand? . . . You and your tape recorder . . .

Yes, I do defend it! I fucking do defend Soviet power. Our power. The people's power! When we had Soviet power we were strong, everyone was afraid of us. The whole world was watching. Some folks shaking with fear and some folks jealous, dammit! And what've we got today? Now? Under democracy? Snickers, and rancid margarine they pass off on us, out-of-date medicines and worn jeans – treating us like fucking natives just climbed out of the trees. Fucking palm trees. Makes you 'wince for our once great power'! Don't you get it? We're up shit creek . . . What a power we were, dammit! Till Gorbachev flew up to his perch, to his throne. The Devil with a birthmark! 'Gorby'. Gorby acted out their plans, the CIA's plans . . . What are you trying to tell me? Get it into your head . . . They blew up Chernobyl, the CIA lot and the democrats. I read it in the paper . . . If Chernobyl hadn't exploded, our great power would never have collapsed. A great power, dammit! (*More expletives.*) Get it into your head . . . a loaf of bread under the Communists cost twenty kopecks, and now it's 2,000 roubles. I could buy a bottle of vodka for three roubles and still have something for a bite . . . But under the democrats? I've been saving up for two months and still can't afford a new pair of trousers. I'm walking about in a ragged sweater. They've sold everything off! Pawned it! Our grandchildren won't be able to pay our debts . . .

I'm not drunk, I'm for the Communists! They were for us, for ordinary people. I don't want any of your tales. Democracy! They did away with censorship. You can write anything you like. A free man . . . Dammit! Well if this free man dies, there's no money for his funeral. An old woman died here. On her own, no children. Two days she was lying in her cottage, poor old thing, wearing her old cardigan, under the icons . . . They couldn't buy her a coffin . . . A Stakhanov shock-worker she was at one time, a team leader. We didn't go out to the fields for two days. We had a protest meeting, dammit! Until the collective-farm chairman came and talked to us, in front of the people, and promised that now, when someone dies, the farm will pay for a wooden coffin, and a *truna* – that's our word for a calf or a piglet – and two crates of vodka for the wake. Under the democrats! Two crates of vodka . . . free! One bottle per man is a piss-up, half a bottle is medicine. For us, from the radiation . . .

Why aren't you recording this? What I'm saying. You only record what's going to make you money. Destabilizing people, rebelling . . . Need some political capital? So you can stuff your pockets full of dollars? We live here . . . suffering . . . And it's nobody's fault! Tell me whose fault it is! I'm for the Communists! They'll be back, and then they'll soon enough find the guilty ones. Dammit! Get it into your head, you people come here, recording . . .

Oh, fuck it all . . .

Declined to give his name

Monologue on how two angels took little Olenka

I've got material. All my bookshelves at home are full of large folders. I know so much I can't write any more . . .

I've been collecting it for seven years: newspaper cuttings, secret instructions, leaflets, my own notes . . . I've got statistics. I'll give you it all. I can fight: organize demonstrations, pickets, get hold of medicine, visit sick children; but I can't write. You do it. I have so many emotions I can't cope with them, they paralyse me, stop me from writing. Chernobyl already has enough guides, enough writers . . . and I don't want to join those people milking the topic. You have to write honestly. Write about everything . . . (*Becomes pensive.*)

Warm April rain . . . For seven years, I've been remembering that rain. The raindrops rolled about like mercury. They say radiation is colourless, but the pools were green or bright yellow. A neighbour whispered to me there had been something on Radio Liberty about an accident at the Chernobyl atomic power station. I didn't pay the slightest attention. I had absolute confidence that if there was anything serious there would be an announcement. We have special equipment, a special alarm system, there are air-raid shelters. We would be warned. We were certain of that! We had all been on civil defence courses. I had even taught on them, been an examiner. That evening, my neighbour brought pills of some sort. A relative had given her them and explained how to take them: he worked at the Institute of Atomic Physics, but made her promise not to tell anyone where they had come from. She was to be as mute as a fish! As a rock! He was particularly anxious about people talking or asking questions on the phone.

I had my little grandson living with me at the time. So what did I do? I still didn't believe a word of it. I don't think any of us took those pills. We were very trusting. Not only the older generation, young people too.

I remember my first impressions, the first rumours. I pass from one time to another, from one emotional state to another. From here back to there . . . As a writer, I've thought about these transitions. They interest me. It's as if there are two people, the pre- and post-Chernobyl me. Only it's difficult to reconstitute the pre-Chernobyl me convincingly. My outlook has changed . . .

I went into the Zone from the very beginning. I remember stopping in a village and being struck by the silence. No birds, nothing. You walk down a street . . . silence. Well, of course, I knew all the cottages were lifeless, that there were no people because they had all left, but everything around had fallen silent. Not a single bird. It was the first time I had ever seen a land without birds, without mosquitoes. Nothing flying in the air.

We went to a village called Chudyany. It registered 150 curies. In Malinovka, the reading was fifty-nine. The population had been exposed to doses hundreds of times greater than those of soldiers guarding the test sites for atomic weapons. Nuclear test sites. Hundreds of times more! The radiation meter would be clicking away, going off the scale, while in the collective-farm offices we found notices signed by the district radiologists saying it was fine to eat onions, lettuce, tomatoes and cucumbers. It all grew there, and everybody ate it.

What do those district radiologists have to say for themselves now? Those secretaries of district Communist Party committees? What excuses can they find?

In all the villages, we met a lot of drunk people. Even women were tipsy, especially the milkmaids and those who looked after the calves. They were singing a song from the movie *The Diamond Arm*, which was popular at the time: 'We don't care, we don't care . . .' In short, nothing really matters.

In Malinovka, in Cherikov District, we visited a kindergarten. The children were running around in the yard. The little ones were crawling about in the sandpit. The head told us the sand was changed every month. She didn't know where it was brought from. You can imagine! The children were all sad. We tried joking with them, but they didn't smile. Their carer was crying. She said, 'Don't. Our children never smile. They even cry in their sleep.' We met a woman with a new baby in the street. 'Who allowed you to give birth here? The radiation is fifty-nine curies.' 'The radiology doctor came,' she said, 'and advised me not to dry nappies outside.' People were urged not to leave. A workforce was needed! Even when villages were evacuated permanently, they still brought people in to work the fields, to dig up the potatoes.

What have they to say now, those secretaries of district and provincial committees? What excuses do they have? Which of them were to blame?

I saved a lot of official instructions. 'Top secret'. I'll give you the lot. Instructions on processing contaminated chicken carcases. In the workshop processing them, you had to be dressed as if you were in contaminated territory and in contact with radioactive elements: rubber gloves, rubber coats, boots and so on. If there were so-and-so many curies, the carcases had to be boiled in salted water, the water drained into the sewer, and the meat could then be added to pâtés and sausage. If the contamination was so-and-so many curies, the meat could be added to bonemeal and used as feed for livestock. That is how they fulfilled the Plan for meat production. Calves from contaminated districts were sold cheaply in other, clean, places. The drivers who drove these calves around told me they looked funny. Their hair reached down to the ground, and they were so hungry they would eat anything, even rags and paper. It was very easy to feed them! They sold them to collective farms, but if anybody wanted to, they could take one home. To their own farm. Criminal goings-on! Criminal!

We came across a truck on the road. It was moving at a funereal pace. We stopped it. There was a young fellow at the wheel. I asked him: 'Is it because you're feeling ill that you're driving so slowly?' 'No,' he said, 'I've got a load of radioactive soil.' On a hot day! With all that dust! 'You must be crazy! You'll want to get married and have children.' 'Well, where else am I going to earn fifty roubles for a single trip?' At that time, fifty roubles would have bought him a good suit. People talked more about bonuses than radiation. Bonuses and various meagre allowances. Meagre in terms of the value of a life.

Tragedy and comedy went side by side.

Some old women were sitting on the benches outside a cottage. Children were running about. We measured seventy curies.

'Where are the children from?'

'They've come from Minsk for the summer.'

'But you have very high radiation levels!'

'Why do you make such a fuss and bother about Radiation! We've seen her.'

'You can't see it!'

'Just you look over there, dearie, to where that cottage is standing half-built. The people upped and left. Got so fearful, they did. Anyway, we went there one evening to have a look. We look in through the window and there she is sitting under a beam, this Radiation of yours. Nasty bit of work, if you ask me, her eyes all shining. And as black as can be . . .'

'That's impossible!'

'We'll swear to it, dearie. Cross our hearts and hope to die!'

Which they did, with a lot of merriment. We weren't sure if they were laughing at themselves or at us.

After our trips, we would meet up again at the editorial office. 'How did you get on?' we would ask each other. 'Oh, fine!' 'Fine? Take a look at yourself in the mirror: your hair's gone grey!' There were jokes. Chernobyl jokes. The shortest one was, 'A good people the Belarusians were.'

I was given an assignment to write about the evacuation. In Polesye, they have a belief that you should plant a tree before going on a long journey, if you want to be sure to return home. I arrived. Went into one yard, then another. Everyone was planting trees. I went into a third yard, sat down and cried. The lady who owned it pointed out to me: 'My daughter and son-in-law planted a plum tree; my second daughter, a black rowan; my eldest son, a viburnum, and the youngest, a willow. I and my old gentleman planted one apple tree between us.' When we parted, she said, 'I have so many strawberry plants, the yard is full of them. Take some of my strawberry plants.' She so wanted something to remain, some trace of her life.

I didn't get round to writing much down. Kept putting things off, thinking I'd sit down sometime and recollect it all. Go off on a break.

Oh! Here's something that's just come back to me. A village graveyard, a sign on the gates saying: 'High radiation levels. Entry by vehicle or on foot prohibited.' You're not even allowed through the Pearly Gates. (Laughs suddenly, for the first time in this long conversation.)

Have people told you that taking pictures anywhere near the reactor was strictly forbidden? You could only do it with a special permit. They confiscated cameras. Before they could leave, the soldiers who had served there were searched, just like in the Afghan War, to make sure they had no photos. God forbid! No evidence. They took the television crews' footage off to the KGB. Returned it after they'd exposed it to light. The amount of information they destroyed! Testimony. All lost to science. To history. It would be good to find the people who ordered that . . .

How would they try to explain themselves? What could they come up with?

I will never forgive them. Never! Because of a little girl . . . She danced in the hospital, danced a polka for me. It was her ninth birthday. She danced so prettily. Two months later, her mother phoned to say, 'Olenka is dying!' I couldn't bring myself to go to the hospital that day, and afterwards it was too late. Olenka had a younger sister. She woke up one morning and said, 'Mummy, I had a dream and I saw two angels coming to take our Olenka. They said Olenka would be happy there. She wouldn't have anything hurting. Mummy, two angels took Olenka . . .'

I can't forgive any of them . . .

Irina Kiselyova, journalist

Monologue on the unaccountable power of one person over another

I am not an arts person, I am a physicist. So facts, just the facts . . .

Somebody will have to answer for Chernobyl. The time will come when people will have to answer for it, as they have been brought to book for the 1937 Purges. Even if it's not for another fifty years! Even if they are very old, even if they are dead. They will answer for it, because they are criminals! (*After a pause.*) It's important to preserve the facts, the facts! Because they will be needed.

On that day, 26 April 1986, I was in Moscow on business. That was where I heard about the accident.

I tried to phone Minsk, to Slyunkov, the first secretary of the Byelorussian Central Committee. I rang one, two, three times, but they wouldn't put me through. I found his assistant (who knew me well): 'I'm calling from Moscow. Put me through to Slyunkov, I have urgent information for him. About the accident!'

I was phoning on the government network, but they had already classified everything. The moment you started talking about the accident, the phone went dead. They had you under surveillance, of course they did! The phone was tapped. The 'appropriate authorities', the government within the government. And this despite the fact that I was phoning the first secretary of the Central Committee . . . And who was I? The director of the Institute of Atomic Energy of the Byelorussian Academy of Sciences. A professor, a corresponding member of the Academy of Sciences, and this information was being kept secret from me.

It must have taken two hours before Slyunkov himself picked up the phone. I reported: 'This accident is serious. According to my calculations' (and I had already talked to some people in Moscow and worked one or two things out), 'a column of radioactivity is moving towards us. Towards Byelorussia. It is essential to carry out urgent iodine prophylaxis for the population and to evacuate everybody living near the power station. People and animals must be moved 100 kilometres away.'

'I've already received the report,' Slyunkov replied. 'There was a fire there, but it has been put out.'

I lost my temper. 'They're lying! They are manifestly lying! Any physicist will tell you that graphite burns at a rate of five tonnes an hour. Imagine how long it will continue burning!'

I took the first train to Minsk. A sleepless night. In the morning, I was home. I took a reading of my son's thyroid gland: 180 microroentgens an hour! At that moment, the thyroid gland was a perfect radiation monitor. What was needed was potassium iodide, standard iodine. Two or three drops in half a glass of fruit jelly for children, and three or four drops for adults. The reactor was burning for ten days, and for ten days people should have been taking that. Nobody would listen to us scientists and doctors. Medicine and science were being dragged into politics. Of course they were! You mustn't forget the intellectual context in which all this was happening, the kind of people we were at that time, ten years ago. The KGB were on the loose, the secret services. The Western radio broadcasts were being jammed. There were thousands of taboos, Party and military secrets. Secret guidelines. And to cap it all, we had been brought up to believe that the peaceful Soviet atom was as harmless as peat and coal. We were fettered by fear and prejudice. Misplaced faith. But facts, let's just keep to the facts.

That same day, 27 April, I decided to go to Gomel Province, on the border with the Ukraine, to the district centres of Bragin, Khoyniki and Narovlya, which are only a few dozen kilometres away from the power station. I wanted the fullest possible information, needed to take instruments, to measure the background radiation. And the background radiation I found was: in Bragin, 30,000 milliroentgens per hour; in Narovlya, 28,000. People were sowing the fields, ploughing. They were preparing for Easter, painting eggs, baking Easter cakes. 'Radiation? What's that? We've had no orders.' People were getting demands from their superiors for reports on how the spring sowing was going: was it proceeding at the required rate? They stared at me as if I were mad, 'Where from? What are you talking about, professor?' Roentgens, microroentgens? It was a language spoken on a different planet.

We returned to Minsk. On the Prospect, there was the usual brisk trade selling pies, ice cream, mince, buns. Under a radioactive cloud.

On 29 April – I remember everything exactly, all these dates – at eight in the morning, I was already sitting in Slyunkov's waiting room. I tried and tried to get in, but nobody was letting me near him. This went on until half past five that evening. At half past five, one of our well-known poets with whom I was acquainted emerged from Slyunkov's office.

'I have been discussing some issues in Belarusian culture with Comrade Slyunkov.'

'There soon won't be anybody to carry it on,' I exploded, 'or to read your books if we don't evacuate people from Chernobyl right now! We have to save them!'

'Oh, come now . . . The fire there has already been extinguished.'

I finally forced my way through to Slyunkov. I described the situation I had found the day before. I told him it was essential to save people's lives! In the Ukraine (I had already phoned there), they had organized an evacuation.

'What are you doing, letting your radiation monitoring technicians' (from my institute) 'run all over the city spreading panic? I've consulted Moscow, spoken to Academician Ilyin. Everything here is fine. The army has been sent into the breach, military equipment . . . A government commission is working at the power station, the State Prosecutor's Office. They're investigating everything. We must not lose sight of the fact that there is a Cold War being waged. We are surrounded by enemies . . .'

Our land already had thousands of tonnes of caesium lying on it, iodine, lead, zirconium, cadmium, beryllium, boron, an unknown quantity of plutonium (the Chernobyl-style uranium-graphite high pressure tube reactors were used to produce weapons-grade plutonium, needed in the manufacture of atomic bombs). In all, 450 types of radionuclides. The quantity was equivalent to 350 bombs of the type dropped on Hiroshima. The authorities needed to be talking about physics, the laws of physics, but they were talking about enemies. Looking for enemies.

Sooner or later, these people will have to answer for what they did. 'At some future time, you will try to explain yourself on the grounds that you were only a tractor manufacturer,' I told Slyunkov (who had previously been the director of a tractor factory, and didn't know anything about radiation). 'But I am a physicist, and I can picture the consequences of this.' Well, whatever next? Some professor, some physicists were daring to lay down the law to the Central Committee? Actually, those people were not a gang of crooks. The best way to characterize it is as a conspiracy of ignorance and corporatism. Their guiding principle, their bureaucratic training, had taught them never to show initiative, to be obsequious. And sure enough, Slyunkov was promoted, moved to Moscow. How do you like that? I suppose there was a phone call from the Kremlin, from Gorbachev. 'Now then, you people there in Byelorussia, let's have no panic. The West is already making quite enough fuss.' The rules they play by are that if you don't do as your superiors want, you'll get no promotion, you won't get to the holiday resort you were hoping for, or get that nice dacha in the country. You have to make yourself liked. If we were still a closed system today, behind the Iron Curtain, people would be living right next to the reactor itself. It would all have been classified! Remember how they hushed up the contamination after an accident at the Kyshtym plutonium production plant in 1957, or the effects of the fallout after atomic tests at Semipalatinsk in 1949 and 1951? That was Stalinism. We still are a Stalinist country.

The secret instructions about what to do if there was a threat of nuclear war specified immediately carrying out the precautionary administration of iodine to the population. And that was just if there was a threat! But here we were looking at 3,000 microroentgens per hour. And all they cared about wasn't people, but managing to hold on to power. A country where power matters and people do not. The primacy of the state is unchallengeable, and a human life is without any value at all. There were things that could be done! We suggested, with no announcements, no panic, just adding medicinal iodine to the reservoirs that provided drinking water, adding it to milk. Of course, people would detect an unusual taste in the water. The milk wouldn't taste quite right. We had a reserve of 700 kilograms of medicine in Minsk. It stayed right where it was, in the depots. They were more afraid of the wrath of their superiors than of a nuclear disaster. Everyone was waiting for a phone call, for orders, and nobody did anything on their own initiative. They were terrified of taking personal responsibility. I carried a dosimeter in my briefcase. Why? They would not let me through to see the big bosses in their offices, they were fed up with me. So then I would take out the meter and apply it to the thyroids of their secretaries and personal drivers sitting there in the waiting rooms. That scared them and sometimes got me seen. 'Look, what's all this hysteria about, professor? Do you think you're the only person with the best interests of the Belarusian people at heart? People are going to die of something anyway: from smoking, in car accidents, by committing suicide.' They ridiculed the Ukrainians. They were crawling on their knees to the Kremlin, begging for money, medicine, radiation monitoring equipment (there was a shortage), but our people (i.e. Slyunkov) summarized the situation in fifteen minutes: 'Everything is under control. We can cope on our own.' He was praised for that: 'Well done, brother Byelorussians!'

How many lives were paid for that praise?

I have information that they themselves, the bosses, were taking iodine. When the staff in our institute examined them, they all had normal thyroids. That's impossible unless you are taking iodine. They were also quietly moving their children elsewhere, out of harm's way. When they had to make trips to the Zone, they had respirators and special boiler suits. All the things other people didn't have. And it's been no secret for a long time that there was a special herd near Minsk where each cow had a number and was assigned to a particular individual. His personal cow. Special land, special greenhouses, special monitoring. The most disgusting thing . . . (*Pauses.*) . . . is that nobody has yet been called to account for that.

Officials stopped allowing me into their offices, stopped listening to me. I started bombarding them with letters and memoranda. I distributed maps, statistics, to every institution. They added up to four folders, each holding 250 sheets of paper. Facts, only facts. Just to be on the safe side, I always made two copies. One was kept in my office at work, and the second was hidden at home. My wife hid them. Why did I make copies? Because we needed to remember this. I know the kind of country we live in . . . I always locked up my office personally. I came back from one work trip to find the files had disappeared. All four thick folders . . . But I was born and bred in Ukraine, my grandparents were Cossacks. Cossack obstinacy. I carried on writing, giving talks. It's our duty to save people's lives! They must be evacuated as a matter of urgency! We were constantly travelling there. Our institute drew up the first map of contaminated areas. The whole of the south was coloured red. The south was ablaze.

Now that's all history. The history of a crime.

All the institute's equipment for monitoring radiation was removed. Confiscated. Without explanation. I received threatening phone calls at home: 'Stop trying to scare people, professor! We'll send you off somewhere you'd really rather not be. Don't know where we mean? Have you forgotten? Don't be in such a rush to forget!' The institute's staff were bullied. Intimidated.

I wrote to Moscow . . .

I was summoned by Platonov, the president of our Academy of Sciences:

'Some day, the Belarusian people will remember you. You've done a lot for them, but it is a pity you wrote that letter to Moscow. A great pity! They're demanding that I dismiss you. Why did you write it? Do you not know whom you are challenging?'

I had maps, statistics. What did they have? They could lock me up in a lunatic asylum. That was what they were threatening. I might die in a car accident. They warned me of that too. They might bring charges against me. For anti-Soviet activity. Or for taking a box of nails not recorded by the institute's supplies manager.

They did bring criminal charges against me.

They got what they wanted. I suffered a heart attack ... (*He is silent.*)

There is everything in those folders. Facts and figures. Criminal statistics.

In the first year after the disaster, millions of tonnes of contaminated grain were processed as animal feed, given to cattle, and the meat then found its way on to our tables. Poultry and pigs were fed bonemeal laced with strontium.

They evacuated villages, but carried on sowing the fields. Our institute's data indicated that one third of the land of collective and state farms was contaminated with caesium-137, and frequently the level exceeded fifteen curies per square kilometre. There was no possibility that it could produce clean food: you couldn't even stay on it for any length of time. Strontium-90 had settled on many fields.

In the villages, people were feeding themselves from their vegetable plots, and there was no monitoring. No one explained to them, or taught them the new ways in which we now needed to live. There was not even an attempt to do so. All that was monitored was what was being sent outside, the state procurement deliveries to Moscow, to Russia.

We did spot checks on children in the villages, several thousand boys and girls. They had readings of 1,500, 2,000, 3,000 microroentgens. Over 3,000. These girls are never going to have healthy babies. They have genetic markers.

So many years have passed, but sometimes I wake up and can't get back sleep.

A tractor is ploughing ... I ask the official from the District Party Committee who's accompanying us, 'Is that tractor driver protected? Does he at least have a respirator?'

'No, they don't wear respirators while they are working.'

'What, have they not brought you any?'

'Of course we have! We've got enough to see us through to the year 2000, but we don't issue them. It would start a panic. They would all run away, go to other regions!'

'What kind of way is that to treat people?'

'It's easy for you to talk like that, professor. If you get fired, you'll just find another job. But I would be completely stuck.'

What sort of a regime is that? Unaccountable power of one person over another . . . This was not just deceit, this was a massacre of the innocents.

Along the River Pripyat, tents, people holidaying with their families. Swimming, sunbathing. They have no idea that for several weeks they've been swimming and sunbathing under a cloud of radioactivity. We've been strictly forbidden to talk to them, but I see these children . . . I go over and start explaining to them. Surprise. Bewilderment. 'Why are they not saying anything about this on the radio and television?' We have an escort. A representative of the local authority, somebody from the District Committee, usually accompanied us. That was the rule. He said nothing. I could tell from his face the conflict of emotions going on inside him: should he report it or not? He was, after all, sorry for these people! He was a normal human being. I didn't know which of his emotions would win out when we returned. Would he denounce me? Everyone made their own choice at that time. (*Says nothing for a while.*)

We are still a Stalinist country . . . Stalin's kind of person is still alive.

I remember, in Kiev, at the railway station. Trains, one after another, taking away thousands of frightened children. Men and women crying. For the first time, I thought, 'Who needs this kind of physics, this kind of science, at such a high price?' Now it's all out in the open. They've written about the amazing shock-working tempos at which the Chernobyl nuclear power plant was built. It was built the Soviet way. The Japanese take twelve years to develop a facility like that, but we did it in just two or three. The quality and reliability of that highly complex facility was what you might expect in an animal-breeding complex, a chicken farm! If there was a shortage of something, they just ignored the plans and substituted whatever was to hand at the time. Thus the roof of the turbine hall was covered with bitumen. That's what the firemen extinguished. And who was in charge of this atomic power station? There wasn't a single nuclear physicist in the management team. They had power engineers, turbine specialists, political workers, but not a single expert. Not a single physicist.

Man has invented a technology for which he is not yet ready. He is not up to it. Can you put a pistol in the hands of a child? We are reckless children. But I'm being emotional, and that is something I forbid myself.

There are radionuclides on the land, in the ground, in the water. Dozens of radionuclides. We need radioecologists, but there are none in Belarus. They had to be called in from Moscow. At one time, we had a Professor Cherkasova working in our Academy of Sciences. Her research was into the effect of small doses, internal radiation. Five years before Chernobyl, her laboratory was closed. 'We are never going to have disasters in the Soviet Union. What are you even suggesting? Soviet atomic power plants are advanced, the best in the world. What small doses are you talking about? What is internal radiation? Radioactive foodstuffs indeed!' They wound down the laboratory and pensioned off the professor. She got a job as a cloakroom attendant somewhere, looking after people's hats and coats.

And nobody has been called to account for anything.

Five years later, the incidence of thyroid cancer in children had increased thirty-fold. An increase was observed in congenital malformations, renal and cardiac disease, paediatric diabetes.

Ten years later, the life expectancy of Belarusians had fallen to sixty years.

I believe in history, the judgement of history . . . Chernobyl is not over. It has only just started.

Vasily Borisovich Nesterenko, former director of the Institute of Atomic Energy, Belarus Academy of Sciences

Monologue on sacrificial victims and priests

A person gets up early in the morning, starts the day . . .

He's not thinking about eternity; his thoughts are on earning his daily bread. But you humanists want to make people think about eternity. It's a mistake you all make.

What is Chernobyl?

We drive into a village. We have a German minibus, donated to our foundation. Children throng around. 'Miss! Sir! We're Chernobyl children. What have you brought us? Give us something. Please!'

That is Chernobyl.

On the road to the Zone, we come across an old woman wearing an embroidered skirt and apron, our folk costume, with a bundle on her back.

'Where are you going, Grandma? Visiting someone?'

'I'm going to Marki, to my home.'

The radiation there is 140 curies! She has twenty-five kilometres to walk each way. It takes her one day there, and another day back. She may retrieve a three-litre jar that's been hanging on her fence for two years, but at least she will have been back home.

That is Chernobyl.

What do I remember of those first days? What was it like? Well, for that we have to go back . . . If you're going to tell the story of your life, you have to begin with childhood. It's the same here. I have my own personal starting point. I start from what seems like a quite different story. I remember the fortieth anniversary of victory in the Second World War. That was the first time we had a fireworks display in Mogilyov. After the official celebrations, people didn't disperse as usual but began singing songs. It was completely unexpected. I remember that shared feeling. After forty years, everybody started talking about the war and thinking about it properly. Before that, we had just been aiming to survive, rebuilding, having children. It will be the same with Chernobyl. We will come back to it later and understand it more deeply. It will become a place of pilgrimage, a Wailing Wall. For the present, though, we have no ritual, no ideas! Curies, rems, sieverts – that doesn't add up to understanding. It's not a philosophy, not an outlook. In this country, man comes either with a gun or with the Cross. That goes right through our history. There has been nothing else. There still isn't.

My mother worked at the headquarters of the city's civil defence team. She was one of the first to hear. All the alarms went off. The instructions they had hanging in every office required them to inform the population immediately, issue respirators, gas masks and so on. They opened up their secret stores, behind all the seals and sealing wax, and found everything in a terrible state. It had deteriorated and was quite useless. In schools, the gas masks were pre-war models, and in any case did not fit children. Their radiation meters were going off the scale, but nobody could understand what was happening. This was completely unprecedented. They simply turned their instruments off. Mother explained, 'If it had been war, we would have known what to do. There were instructions. But here?' Who was in charge of civil defence? Retired generals, colonels for whom a war began with government announcements on the radio, followed by air-raid warnings, bombs, incendiaries. It hadn't dawned on them that these were different times. It needed a leap of the imagination, which did eventually happen. We know now that people will be sitting, drinking tea at the weekend . . . We will talk and laugh, and meanwhile a war will be going on. We won't even notice when we disappear.

Civil defence was just a game played by grown men. Their responsibility was to hold parades and exercises. It cost millions of roubles. We were dragged away from work for three days, without any explanation, for military exercises. The game was called 'In Case of Nuclear War'. The men played soldiers and firemen, and the women were nurses. We were issued with boiler suits, boots, first-aid kits, as well as a pack of bandages and medicines of some description. Everything was just as it should be. The Soviet people must face the enemy worthily. Secret maps and evacuation plans were stored away in fireproof safes with wax seals. According to these plans, the populace were to be organized and taken, within a matter of minutes after the alarm, into the forest to a place of safety. A siren wails. Attention, attention! We are at war . . .

Cups and banners were awarded, and there was a feast. The men drank to our coming victory and, of course, to the women!

Only the other day, the alarm was sounded in the city. Attention! Civil defence! That was a week ago. People were frightened, but it was a different kind of fear. This was not an attack by the Americans or Germans. Was something going on at Chernobyl? Had it really happened again?

1986. Who were we? What were we when this technological version of the end of the world came upon us? I? We? The local intelligentsia? We had our own circle. We lived separate lives, withdrawing from everything around us. It was our way of protesting. We had our own laws: we didn't read *Pravda*. Instead *Ogonyok* was passed from hand to hand. The instant the reins were loosened, we revelled in the new freedom. We read samizdat, which finally reached us in our backwater. We read Solzhenitsyn and Shalamov, Venedikt Yerofeyev. We met at each other's homes and had endless discussions in the kitchen. We had this great yearning. For what? Somewhere, we were sure, there was a world inhabited by actors and film stars. I so wanted to be Catherine Deneuve, to wear some ridiculous Greek mantle and curl my hair in an improbable manner. It was a yearning for freedom. There was a world out there about which we knew nothing, a foreign world, a manifestation of freedom. But this too was a game, a flight from reality. Some members of our circle gave in, became alcoholics, or joined the Party to crawl up the career ladder. Nobody believed it was possible to breach the Kremlin wall, to break through and make it collapse. Not in our lifetime, anyway, that was for sure. That being so, to hell with what they were getting up to in there. We would live our lives here, in our world of illusion.

Chernobyl. At first, we had the same reaction. What has it got to do with us? Let the regime worry about it. Chernobyl was all their doing and it was a long way away. We didn't even bother to look it up on the map. Not interested. We no longer felt any need for the truth. It was when different labels appeared on the milk bottles: 'Milk for children' and 'Milk for adults', that we thought, 'Hello! Something is coming too close to home.' I was no Party member, but I was still Soviet. You began to feel alarmed: 'The leaves on radishes this year look more like beetroot tops.' But then, in the evening, you would turn on the television and hear, 'Do not fall for Western provocation!' and all your doubts were dispelled. The May Day Parade? Nobody forced us to turn out for it; nobody obliged me to. We could've decided not to go, but we didn't. I don't remember a more crowded, happy May Day Parade than the one that year. There were concerns, of course, but you wanted to run with the herd, to feel the sense of solidarity, to be in there together with everybody else. I wanted someone to blame. The bosses, the government, the Communists. Now I look back, trying to find the moment something broke. Where did it break? Well, actually, right at the beginning. Our lack of freedom . . . the height of dissidence was to question whether or not you could eat the radishes. The servility was inside us.

I was working as an engineer at the synthetic-fibre factory, and we had a group of German experts staying with us. They were installing new machinery. I saw the way other people behaved, a different nation, people from another world. When they heard about the accident, they immediately demanded doctors, radiation meters, and to have their food tested. They listened to their own radio broadcasts and knew what action needed to be taken. Of course, they weren't issued a thing. Then they packed their bags and got ready to leave. 'Buy tickets for us! Send us home! We're leaving, since you're incapable of ensuring our safety.' They went on strike, they sent a telegram to their government, to their president. They fought for their wives and children, who were living here with them. For their lives! And how did we view their behaviour? Oh, just look at those Germans, they're hysterical! Cowards! Measuring the radiation in their borscht, in their meatballs! Trying not to go outside more often than necessary. What clowns! Look at our men, they're real men! That's Russians for you! Daredevils! Battling the reactor! They're not afraid for their own skin! Climbing up on to the molten roof with their bare hands or in canvas gloves (we'd seen them on television). And our children were out there with their flags at the May Day Parade! And the war veterans, the old guard! (*Becomes pensive.*) But that is a kind of barbarism too, the lack of a sense of self-preservation. We always say 'we' instead of 'I'. 'We will show what Soviet heroism is!' 'We will show the character of the Soviet people.' To the whole world! But then what about me? I don't want to die either. I'm frightened.

It's interesting today to look back at myself and my feelings, how they were changing. To analyse them. I long ago caught myself paying more attention to the world around. Around me and within me. After Chernobyl, that happened automatically. We began learning to say 'I'. 'I don't want to die!' 'I'm afraid' . . . And then? I turn the television up: a red banner is being presented to milkmaids who have been 'victors in Socialist competition'. But that's here! Near Mogilyov! In a village which is in the centre of a caesium hotspot! The people there are about to be resettled, any time now. The presenter intones, 'People are working selflessly, no matter what . . . Miracles of courage and heroism.' Never mind if it's a new Flood! 'Forward march, Children of the Revolution!' You're right, I'm not a member of the Party, but I am still a Soviet citizen. 'Comrades, do not be fooled by provocateurs!' the television booms night and day. Our doubts are dispelled. (*A telephone call. Returns to the conversation half an hour later.*)

I'm interested in every new person. Everyone who thinks about this . . .

Sometime in the future, we will understand Chernobyl as a philosophy. Two states divided by barbed wire: one, the Zone itself; the other, everywhere else. People have hung white towels on the rotting stakes around the Zone, as if they were crucifixes. It's a custom here. People go there as if to a graveyard. A post-technological world. Time has gone backwards. What is buried there is not only their home but a whole epoch. An epoch of faith. In science! In an ideal of social justice! A great empire came apart at the seams, collapsed. First there was Afghanistan, then Chernobyl. When the empire disintegrated, we were on our own. I hesitate to say it,

but . . . we love Chernobyl. We have come to love it. It is the meaning of our lives, which we have found again, the meaning of our
suffering. Like the war. The world heard about us Belarusians after
Chernobyl. It was our introduction to Europe. We are simultaneously its sacrificial victims and its priests. That is a terrible thing to
say. I realized it only recently.

In the Zone itself, even the sounds are different. You go into a
house and expect to come across Sleeping Beauty. If it has not yet
been looted, there are photographs, pots and pans, furniture . . . You
feel the people who lived here must be somewhere nearby. Sometimes we do find them, but they don't talk about Chernobyl. They
tell you they have been cheated. Their big worry is whether they
will get everything they are entitled to, and whether someone else
might get more. Our people have always felt they were being
cheated, at every stage of their history. On the one hand, there is the
nihilism, the rejection, and on the other, fatalism. They have no
faith in the government. They don't trust scientists and doctors, but
they don't do anything for themselves. They are innocent and
vacant. They have found the meaning and justification of their
existence in suffering. Nothing else seems to matter. Along the sides
of fields are signs saying 'High radiation levels', but the fields are
being ploughed. Thirty curies, fifty . . . The tractor drivers are
sitting in open cabins breathing radioactive dust. Ten years have
passed, yet we still have no tractors with sealed cabins. Ten years!
Who are we, for heaven's sake? We live on contaminated land,
plough, sow, have children . . . What sense is there in our suffering?
What is it for? Why is there so much of it? I talk about that a lot now
with my friends. We discuss it often. Because the Zone is not just
rems and curies and microroentgens. It is our people. Our nation.
Chernobyl rescued our system, just as it was dying. We have gone
back to eternal states of emergency, central allocation of jobs,
rations. They used to drum into our heads, 'If it had not been for the
war . . .' Now they can blame everything on Chernobyl. 'If it had not
been for Chernobyl . . .' Our eyes immediately mist over as we
relapse into self-pity. Give us aid! Give us aid, so we have something
to divvy up between us! The trough. Something to divert our anger!

Chernobyl is already history, but it is also my job, my everyday routine. I travel. I see things. There used to be the patriarchal Belarusian village, the Belarusian cottage. No inside toilet or hot water, but it had an icon, a well with a carved wooden covering, embroidered towels and bedspreads. And hospitality. We went into one cottage for a drink of water, and the woman pulled a towel out of a coffer as old as herself and gave it to me. 'This is for you, to remember your visit to my home.' There was a wood, a field. The commune had survived and some fragments of freedom: her own land beside the house, a smallholding, her own cow. They were being resettled from Chernobyl to 'Europe', European-style settlements. You can build a house that's better and more comfortable, but in a new place you cannot rebuild this vast world to which they were umbilically connected. It's a terrible blow to a person's sense of identity. Uprooting traditions and an age-old culture. As you approach these new settlements, they're like a mirage on the horizon. Painted light blue, dark blue, orangey-red. And the names they give them! Maytime, Sunnyside! European houses are far more convenient than the old cottages. They are the future served on a plate. But you cannot parachute people into the future. The people have been turned into exotic natives. They sit on the ground and wait for a plane or a bus to arrive and bring them foreign aid. There's no sense of being pleased at the opportunity to start over again. 'I've got myself out of the inferno; I have a house, uncontaminated land; I need to work for the future of my children, who have Chernobyl in their blood, in their genes.' They're just waiting for a miracle to happen. They go to church. Do you know what they ask God for? The same thing – a miracle. Not for Him to grant them good health and the strength to achieve something for themselves. They're used to begging now, if not from foreigners then from heaven.

They live in these tidy houses and it's as if they were hutches. They crumble, fall to pieces. The people living there are not free, they are doomed. They live in fear and resentment, and wouldn't hammer in a nail on their own account. They want Communism.

They're waiting for it. The Zone needs Communism . . . In every election, the people there vote for the 'firm hand' candidate. They're nostalgic for the order of Stalin's time, military order. For them, that is synonymous with justice. They even live in military surroundings, with police outposts, people in military uniform, a system of permits, rations. Officials distribute the foreign aid. In German and Russian, the boxes have written on them: 'Not for trade. Not to be sold.' It is sold. You will find it everywhere. In every kiosk.

Again, like a game, a promotional trip, I lead a convoy of humanitarian aid donors. Outsiders, foreigners . . . They come to us in the name of Jesus and in the name of who knows what else. And there are my fellow tribesmen, standing in the puddles and the mud, in jerseys and quilted jackets, in cheap boots . . . 'We don't need anything! The officials will only steal it!' I read in their eyes. But immediately, alongside that, the urge to grab a box, a crate of something foreign. We know by now what kind of old woman lives where. Like in a zoo. It's humiliating, and I feel a wicked, crazy urge. I sometimes suddenly say to them: 'Now we're going to show you something you could never see, even in Africa. There's nothing like it anywhere else in the world! Two hundred curies, 300 curies . . .' I notice too how the old women themselves have changed. Some have become real film stars. They've memorized their lines, and will shed a tear in just the right places. When the first foreigners came, they used to say nothing, just weep. Now, though, they've learned to talk. With a bit of luck, there'll be some chewing gum for the children, or a carton of clothing might come their way. Who knows? And this coexists with a philosophical profundity, with the fact that they have a special relationship here with death and time. They refuse to abandon their cottages, their families' graveyards; and that's not because of German chocolate, or chewing gum.

On our way back, I point out to them: 'See what a beautiful land this is!' The sun has sunk low on the horizon, lighting up the forest and fields, a gift to us as we are leaving. 'Yes,' someone in the

German group who speaks Russian replies, 'beautiful, but poisoned.' He is holding a dosimeter.

I realize that this sunset is dear only to me. This is my native land.

*Natalia Arsenyevna Roslova, chairwoman of the Mogilyov
Women's Committee of Children of Chernobyl*

The Children's Choir

*Alyosha Belsky, aged nine; Anya Bogush, ten; Natasha Dvoretskaya,
sixteen; Lena Zhudro, fifteen; Yura Zhuk, fifteen; Olya Zvonak, ten;
Snezhana Zinevich, sixteen; Ira Kudryacheva, fourteen; Yulya Kasko,
eleven; Vanya Kovarov, twelve; Vadim Krasnosolnyshko, nine; Vasya
Mikulich, fifteen; Anton Nashivankin, fourteen; Marat Tatartsev,
sixteen; Yulya Taraskina, fifteen; Katya Shevchuk, fourteen; Boris
Shkirmankov, sixteen.*

I was in hospital . . .

I was in such pain. I said to my mother: 'Mum, I can't stand it. It's best if you kill me!'

There was this black cloud . . . This really heavy rain . . .

The puddles turned yellow . . . and green . . . as if someone had poured paint in them. People said it was just pollen from the flowers. We didn't run through the puddles, just looked at them. Grandma shut us in the cellar. She knelt down and prayed. She told us to pray too. 'Pray! It's the end of the world. God's punishment for our sins.' My brother was eight and I was six. We started remembering our sins: he had broken a jar of raspberry jam, and I hadn't told Mum my new dress had got caught on the fence and torn . . . I hid it in the wardrobe.

Mum often wears black. A black headscarf. On our street, someone is getting buried all the time. People cry. If I hear music, I run home and pray. I say the Lord's Prayer.

I pray for my mum and dad . . .

*

Soldiers in trucks came to take us away. I thought a war had started . . .

The soldiers had real rifles hanging from their shoulders. They said words I couldn't understand: 'decontamination', 'isotopes' . . . On the journey, I had a dream: there was an explosion, but I was alive! My home had gone, my parents. There were not even any sparrows and crows. I woke up in horror. I jumped up. I peeped through the curtains to see if there was that nightmarish mushroom cloud in the sky.

I remember a soldier chasing a cat. It made the radiation meter rattle like a machine gun: click-click-click. A boy and girl were running after them: it was their cat. The boy didn't say anything, but the girl was shouting, 'You can't have her!' She shouted as she ran, 'Pussy, run for it! Run, pussy!'

The soldier had a big plastic bag.

We left my hamster at home when we locked everything up. He was a little white hamster. We gave him enough food for two days.

But we never went back.

It was the first time I had been on a train . . .

The train was packed with children. The little ones were screaming and dirtying themselves. There was one carer for every twenty children and they were all crying, 'Mummy! Where's my mummy? I want to go home!' I was ten. The girls like me helped to calm the little ones down. Women met us on the train platforms and made the sign of the Cross over the train. They brought us biscuits they had made, and milk and warm potatoes . . .

We were taken to Leningrad Province. There, when we were approaching the stations, people made the sign of the Cross, but watched us from far away. They were afraid of our train, and at every station they washed it down for a long time. When we jumped out at one station and ran to the buffet, they stopped anybody else going in: 'Chernobyl children are eating ice cream in there,' they said. The woman at the counter told someone over the phone, 'When they leave, we'll wash the floor with bleach and boil the glasses.' We heard that.

Some doctors met us. They were wearing gas masks and rubber gloves. They took our clothes and all our things away, even our envelopes and pens and pencils, and put them in plastic bags and buried them in the forest.

We were so scared. For a long time afterwards, we were expecting we would start to die.

Daddy kissed Mummy and I was born . . .

I used to think I would never die, but now I know I will. A boy was lying next to me in the hospital. He was called Vadik Korinkov. He drew little birds for me, little houses. He died. Dying isn't frightening. You just sleep for a long, long time and don't wake up. Vadik told me that, when he died, he would live in another place for a long time. One of the older boys told him. He wasn't frightened.

I dreamed I died. I could hear my mummy crying in my dream and woke up.

We were leaving . . .

I want to tell you how my grandma said goodbye to our house. She asked my dad to bring a sack of millet from the pantry, and scattered it over the garden, 'For God's birds.' She collected eggs in a sieve and scattered them through the farmyard, 'For our cat and dog.' She sliced up pork fat for them. She emptied all the seeds out of her little bags: carrots, pumpkins, cucumbers, her blackseed onions, all the different flowers . . . She shook them out over the vegetable plot: 'Let them live in the soil.' Then she bowed to the house. She bowed to the barn. She went round and bowed to every apple tree.

My grandfather, when we were going away, took his hat off.

I was little . . .

Six, no, eight I think. Yes, eight. I've just counted it. I remember there was so much to be afraid of. I was afraid of running barefoot on the grass. Mum warned me I would die if I did. Swimming, diving – I was scared of everything. Picking nuts in the forest. Lifting up a beetle . . . because he crept over the ground, and it was infected. Ants, butterflies, bumblebees – they were all infected. My mum remembers they told her at the pharmacy to give me a teaspoonful of iodine. Three times a day! But she was too frightened . . .

We were waiting for spring: would the daisies come up again, like they did before? Everybody was saying the world was going to change. On the radio and television . . . The daisies were going to turn into . . . what? Into something else, anyway. And the foxes would grow a second tail. The hedgehogs would be born without prickles, the roses would have no petals. People would appear, and they would be like humanoids and be coloured yellow. They would have no hair, no eyelashes. Only eyes. And the sunsets would be green, not red.

I was little. Eight years old.

Then it was spring. In the spring, leaves unfolded out of the buds like they always did. They were green. The apple trees blossomed. They were white. The cherry trees had the same fragrance. The daisies, they were the same as usual. Then we ran to the river to the anglers, to see if the roach still had heads and tails. And the pike. We checked the starlings' nest boxes to see if they had flown back, and whether they would have babies.

We had a lot of work to do. We had so much to check.

The grown-ups were whispering. I could hear . . .

In the year I was born, 1986, there were no other boys or girls born in our village. I'm the only one. The doctors didn't allow it. They frightened Mum. Something horrid . . . but my mum ran away from the clinic and hid with Grandma. And then there was me . . . just turned up . . . Well, was born, I mean. I overheard all that.

I have no little brother or sister, and I so want one. Where do children come from? I would go and look till I found my little brother myself.

Grandma keeps giving me different answers: 'The stork brings them in its beak. Or sometimes a little girl grows in the fields. Little boys are found in berries, if a bird drops them.'

Mum told me something different. 'You fell from heaven.'

'How?'

'It started raining, and you fell right into my arms.'

Miss, are you a writer? How could there not be a me? And where would I be? Somewhere high up in the sky? Or maybe on another planet?

I used to love going to exhibitions, looking at the pictures . . .

They brought an exhibition about Chernobyl to our town . . . a young colt is running through the forest, but he is all legs, eight or ten of them; a calf has three heads; bald rabbits sitting in a cage, looking as if they're plastic. People walking through a meadow wearing spacesuits. Trees higher than churches, flowers as high as trees . . . I didn't make it to the end. I came upon a photo of a boy stretching out his arms, perhaps for a dandelion, or perhaps to the sun, but instead of a nose he had something like an elephant's trunk. I wanted to cry. I wanted to shout, 'We don't need exhibitions like this! Don't bring them! Everybody around here is already talking about death. About the mutants. I don't want any more!' On the first day it opened, people came, but after that, not a single person. In Moscow and St Petersburg, they wrote all about it in the newspapers and crowds of people went to see it. In our town, the exhibition hall was empty.

I went to Austria for treatment. There are people there who can hang a picture like that up at home – a boy with a trunk, or flippers instead of hands – and look at it every day, in order not to forget about people in misfortune. But when you live here, it's not a fantasy or art, it's real life. My life . . . If I had the choice, I'd rather hang something pretty in my room. A beautiful landscape, where everything is normal, the trees, the birds. Ordinary. Cheerful . . .

I want to think about pretty things.

The first year after the accident . . .

All the sparrows in our village disappeared. They were lying all over the place, in gardens, on the tarmac. They got raked up and taken away in containers, along with the leaves. That year, you weren't allowed to burn leaves. They were radioactive. They had to be buried.

Two years later, the sparrows reappeared. We were so pleased. We were shouting to each other, 'I saw a sparrow yesterday.' They're back.'

The cockchafers vanished. They still haven't come back. Perhaps they will in a hundred or a thousand years' time, like our teacher says. Even I won't see them, though I'm only nine.

And what about my grandma? She's very old.

On 1 September, starting school again . . .

There wasn't a single bunch of flowers. We knew by then there was a lot of radiation in flowers. Before school started, it wasn't carpenters and painters working like it used to be, it was soldiers. They scythed down the flowers, stripped off the soil and took it away somewhere in trucks with trailers. They cut down a big, ancient park, the old lime trees. Old Nadya – she was always called to the house when someone died, to do the keening and say the prayers – said, ''Twas not the lightning struck you . . . Not the drought that brought you low . . . The sea did not flood you . . . Yet there you lie like coffins black.' She mourned the trees as if they were human beings. 'Alas, my oak tree, my apple tree, gone . . .'

A year later, we were all evacuated and they buried the village. My dad is a driver. He took me there, and told me about it. First, they dug a deep pit, five metres deep . . . Then the firemen came with their fire engine and hosed a house down, from the roof ridge to the foundations, so as not to raise radioactive dust. The windows, the roof, the doorway, they washed everything. Then a crane lifted the house up and put it in the pit. Dolls, books, jars everywhere. The digger scooped them up, covered everything with sand and clay and firmed it all down. Where the village had been was just a flat field. Our house was under it, and the school, and the village soviet. My dried-flower collection and two stamp albums. I so wanted to take them with me.

I had a bike. My parents had just bought it for me.

*

I'm twelve . . .

I'm at home all the time, an invalid. In our house, the postman brings a pension for my grandad and one for me. When the girls in my class learned I had leukaemia, they were afraid to sit next to me or touch me. I would look at my hands, my school bag and exercise book. Nothing had changed. Why were they afraid of me?

The doctors said I was ill because my dad had worked at Chernobyl. I was born after that.

But I love my dad.

I had never seen so many soldiers . . .

They washed down the trees, the houses and roofs. They washed down the collective farm's cows. I thought, 'Poor forest animals. Nobody washes them down. They will all die. Nobody washes the forest down. It will die too.'

Our teacher said, 'Draw radiation.' I drew it raining yellow rain, and a red river flowing.

From childhood, I loved technology . . .

I dreamed that, when I grew up, I would work in something technical. My father loved technology. He and I were forever designing something. Constructing.

My dad went away. I didn't hear him leaving. I was asleep. In the morning, I saw Mum all tear-stained. She said, 'Your father is at Chernobyl.'

We waited for Dad to come back, as if he was away in the war.

He came back and went to work at his factory again. He didn't tell me anything. At school, I was boasting to everybody that my dad had come back from Chernobyl; he was a member of the clean-up team, and the clean-up workers were the ones who helped to overcome the accident. Heroes! The other boys envied me.

A year later, Dad became ill . . .

We were walking round the park at the hospital. That was after his second operation. It was the first time he talked to me about Chernobyl.

They were working quite near the reactor. It was as quiet as could be, he said, and beautiful. All sorts of things were happening. Orchards were blossoming, but who for? The people were gone from the villages. They drove through Pripyat. There was washing hanging on the balconies, pots of flowers. A bicycle had been left under a bush with a postman's canvas bag full of newspapers and letters. A bird had nested on it. I saw it all, as if it was a film . . .

They 'decontaminated' things that just needed to be abandoned. They stripped away topsoil poisoned with caesium and strontium, and by the next day it was clicking on the meters all over again.

'When we left, we got a handshake and a certificate of commendation for our selfless work.' My dad reminisced and reminisced. The last time he came out of hospital he told us: 'If I stay alive, I want nothing to do with physics or chemistry. I'll retire from the factory and be a shepherd.'

My mother and I have been left on our own. I am not going to enrol at a technical college, which is what my mother would like. The one where my dad studied.

I have a little brother . . .

He likes playing Chernobyl. He builds air-raid shelters and pours sand on the reactor . . . or else he dresses up as a bogeyman and runs around trying to scare everyone by saying, 'Ooh! I am Radiation! Ooh! I am Radiation!'

He wasn't born when it happened.

At night, I fly . . .

I fly surrounded by bright light. It's not real, but it's not other-worldly either. It is both and neither. In my dream, I know I can go into this world and be in it . . . Perhaps stay in it? My tongue isn't working properly, there's something wrong with my breathing, but there I don't need to be able to talk to anyone. Something similar has happened to me before, but when? I can't remember. I want so much to join everybody else, but I can't see anyone. Only light. I feel I can touch it. How huge I am! I am together with everybody, but somehow aside, apart. Alone. In my very earliest childhood, I saw some coloured pictures like what I am seeing today. In this dream, there comes a moment when I cannot think about anything else. Only, suddenly a window will open. An unexpected gust of wind. What is it? Where has it come from? There is a link now between me and someone else . . . A way to be in touch . . . But how these grey hospital walls hem me in! How weak I still am! I use my head to block off the light because it stops me seeing properly . . . I strain and strain and start looking higher . . .

My mum came. Yesterday, she hung an icon in the ward. She was whispering something there in the corner, knelt down. They're all silent: the professor, the doctors and nurses. They think I have no idea that I'm going to die soon. But at night, I'm learning to fly . . .

Who said flying is easy?

I used to write poetry once . . . I fell in love with a girl in fifth grade. In seventh grade, I discovered there was such a thing as death. My favourite poet is García Lorca. I read his words, 'the dark root of a cry'. At night, poetry has another sound. A different one. I've started learning to fly. It's not a game I like, but what can you do?

My best friend was called Andrey. He had two operations, and then they sent him home. He was supposed to have a third operation in six months' time. He hanged himself with his belt, in an empty classroom, when all the others had run off to the physical education class. The doctors had told him he mustn't run or jump. He had been the best footballer in the school, before the . . . surgery.

I had a lot of friends here: Yulya, Katya, Vadim, Oxana, Oleg . . . and now Andrey. 'We will die and become part of science,' Andrey used to say. 'We will die and everyone will forget us.' That's what Katya thought. 'When I die, don't bury me in a graveyard. I'm afraid of cemeteries. There are only dead people there, and crows. Bury me in open countryside', was what Oxana wanted. 'We are going to die,' Yulya said, and cried.

For me, the sky is alive now when I look up at it. They are up there.

A lone human voice

I was so happy recently. Why? I've forgotten . . .

Everything has been left behind, in some other life. I don't understand. I don't know how I've been able to live again. I suppose I wanted to go on living, and here I am laughing and talking. I missed him so much. It was like being paralysed. I wanted to talk to someone, but not with any other people. I would go into the church, and it would be so quiet there, like in the mountains. So quiet. You can forget about your life there. But in the morning, I wake up and reach out for him. Where is he? His pillow, his smell. A tiny bird I can't identify runs along the windowsill, trilling like a little bell and waking me up. I've never heard a sound, a voice like it. Where is he? I can't convey everything. Not everything can be put into words. I don't know how I managed to go on living. In the evening, my daughter comes to me and says, 'Mum, I've done all my homework.' Then I remember that I have children. But where is he? 'Mum, one of my buttons has come off. Can you sew it back on?' How can I follow after him? See him again? I close my eyes and think about him until I fall asleep. He comes to me in my sleep, but only briefly, fleetingly. He vanishes immediately. I even hear his footsteps. But where does he disappear to? He so much wanted not to die. He looked out the window, at the sky . . . I would prop him up on a pillow, on two, three, to raise him higher. He took so long to die, a whole year. We couldn't bear to part . . . (*A long silence.*)

No, no. It's all right, I'm not going to cry. I've forgotten how to do that. I want to talk . . . Sometimes it's so hard, so unbearable. I try to tell myself, persuade myself, I don't remember anything. Like a friend of mine. Just so as not to go mad. She . . . Our husbands died in the same year. They were together in Chernobyl. She is planning to remarry, wants to forget, to close that door. The door to over there. After him. No, no, I understand her. I know . . . We have to live on . . . She has children . . . We've been somewhere no one else has, seen things no one else has seen. I bottle it up, keep it to myself; but once, in the train, I started telling other people, complete strangers, all about it. What for? It's so terrible being alone.

He went to Chernobyl on my birthday. Our guests were still sitting round the table. He apologized to them, kissed me. There was a car already waiting below our window. It was 19 October 1986. My birthday . . . He was a fitter, travelled the length and breadth of the USSR, and I waited for him. It was like that for years. Our life was the life of people in love, saying goodbye and meeting again. But then . . . It was only our mothers who were gripped by fear, his mother and mine, but he and I were not afraid. I wonder now why that was. We knew very well where he was going. We might at least have borrowed a tenth-grade physics textbook from our neighbour's boy and skimmed through it. He didn't wear a hat there. The other lads in his team had all their hair fall out within a year, but his actually grew thicker. None of them are alive today. His team, seven of them, all died. Young men. One after the other. The first died three years afterwards. Well, we thought, that doesn't mean anything. Just his fate. But then a second died, a third, a fourth . . . Then they were all waiting for it to be their turn. That was how they lived! My husband was the last to die . . . They were fitters, used to heights. They disconnected the electricity in the villages that had been evacuated, climbing poles, clambering over dead houses, working along dead streets. Always at a height, overhead. He was nearly two metres tall, weighed ninety kilograms. Who could kill a big man like him? For a long time, we weren't afraid. (*Suddenly smiles.*)

Oh, how happy I was! When he came back and I saw him again. There was a holiday atmosphere in our home. It was always a celebration when he came back. I have a long, long, very pretty nightie. I would put that on. I loved expensive underwear. All mine is good quality, but this nightdress was special. For special occasions. For our first day together again, our nights . . . I knew every part of his body, intimately. I kissed all of it. I would even sometimes dream I was part of his body, that's how inseparable we were. When he was away, I missed him terribly. It hurt physically. Whenever we parted, I was completely at sea for a while, not sure where I was, what street I was in or what the time was. Time stood still for me.

When he came back, he already had swollen lymph nodes on his neck. I felt them with my lips. They were small, but I asked him, 'Will you go to see the doctor?' He reassured me. 'They'll go away.' 'What was it like there, in Chernobyl?' 'Just normal work.' No bravado, no panic. One thing he did tell me: 'It was just the same there as here.' In the canteen where they had their meals, they served the ordinary people on the ground floor: noodles, tinned food; while on the second, the brass, army generals, they had fruit, red wine, mineral water. Clean tablecloths. Every one of them had a dosimeter for the radiation. They didn't give him a single one for the whole of his team.

I remember the seaside. He and I went to the seaside. What I remember is that there was just as much sea as there was sky. My friend and her husband . . . They came with us too. She remembers: 'The sea was dirty. Everyone was afraid of getting cholera.' The newspapers had been writing something about it. I remember it differently. In very vivid colours. I remember the sea was everywhere, and so was the sky. Such a deep, deep blue. And him there beside me. I was born to love, to be happy in love. At school, all the girls had ambitions. Some wanted to go to college, some to go away on a Young Communist League construction project, but I just wanted to get married. To love incredibly, as passionately as Natasha Rostova in *War and Peace*. Just to love someone! Only I couldn't admit it to anyone, because at that time, you must remember, the only permissible dreams were of going to work on a YCL project. We had that drummed into us. People were desperate to go to Siberia, into the impenetrable forests of the taiga. Do you remember, we used to sing, 'for the mists and for the smell of the taiga'? I didn't manage to get into college at my first attempt. I didn't have enough credits, so I went to work at the telephone exchange. That's where we met. I was on duty. I got him to marry me. I said to him, 'Marry me. I love you so much!' I was head over heels in love. He was so handsome. I was flying in the sky. I asked him myself: 'Marry me!' (*Smiles.*)

At times, I think about things and try to comfort myself in different ways: perhaps death is not the end, and he has just changed and is living somewhere in a different world. Somewhere near me. I work in a library, read a lot of books, meet all sorts of people. I would like to talk about death. To understand. I'm looking for consolation. I find things in the newspapers and in books. I go to the theatre if there's something on there about it, about death. I am physically in pain without him, I can't live on my own . . .

He didn't want to go to the doctor: 'I don't feel anything. I'm not in pain.' By now, his lymph nodes were the size of hens' eggs. I had to push him into the car by force and drive him to the clinic. They referred us to a cancer specialist. One doctor looked at him and called another: 'I've got one more here from Chernobyl.' They kept him in the hospital. A week later they operated: they completely removed his thyroid gland and larynx and replaced them with a lot of tubes. Yes . . . (*Trails off.*) Yes. Now I know that was still a happy time. Lord! What silly things I was doing, running round the shops, buying presents for the doctors: boxes of chocolates, imported liqueurs. Bars of chocolate for the nurses. And they took them. He laughed at me: 'Sweetheart, they're not gods. The chemotherapy and radiation treatment aren't in short supply. They'll give it without expecting chocolates.' But I rushed to the outskirts of the city to get a chocolate and marshmallow cake, or some French perfume. At that time, you could get these things only if you knew someone. It all came from under the counter. That was before they discharged him to come home. We . . . We were coming home! I was given a special syringe and shown how to use it. I had to feed him with it. I learned all that. Four times a day, I cooked something fresh, fresh every time. I ground it up in a mincer, rubbed it through a strainer, and then sucked it up into the syringe. I pierced one of the tubes, the biggest one, and it went into his stomach. But he lost his sense of smell, he couldn't tell one food from another. I would ask, 'Do you like that?' but he didn't know.

In spite of everything, we still managed to get out a few times to the cinema. We would kiss in there. We were hanging by such a very thin thread, but it seemed to us we were managing to hold on to life. We tried not to talk about Chernobyl, never to remember it. It was a taboo subject. I wouldn't let him near the phone. I intercepted it. His lads were dying, one after another. It was taboo . . . But then one morning I woke him, gave him his dressing gown, and he couldn't get up. He couldn't say anything either. He'd stopped talking. His eyes were ever so big. That's when he got really frightened. Yes . . . (*Trails off again.*) He stayed with us for another year. All that year, he was dying. Every day, he got worse and worse, and he knew his mates were dying too. That was something else we had to live with, that waiting. They talk about Chernobyl, they write about it, but nobody knows what it's like. Everything here is different now. We aren't born the same, we don't die the same way. Not like everybody else. Ask me how people die after Chernobyl! The man I loved, loved so much that I couldn't have loved him more if I had given birth to him, turned in front of my eyes into a monster.

They took out his lymph nodes, and that affected his circulation. His nose got somehow out of place and three times bigger, and his eyes weren't the same any more. They moved in opposite directions. There was a different light in them, one I didn't know, and an expression as if it wasn't him looking out of them but someone else. Then one eye closed completely. And what was I really afraid of? I just didn't want him to see himself, to have to remember what he looked like. But he began asking me, showing me with his hands that he wanted me to bring him a mirror. Sometimes I would run out to the kitchen, pretending to have forgotten, or I would come up with something else. I played those tricks on him for two days, and on the third day he wrote in large letters in his notebook, with three exclamation marks, 'Give me a mirror!!!' We already had a notebook, a pen, a pencil, that was how we communicated now because he couldn't speak even in a whisper, he couldn't even manage a whisper. He was completely mute. I ran to the kitchen, rattling saucepans about. I hadn't read it! I hadn't heard! Again, he wrote: 'Give me a mirror!!!' again with all those exclamation marks. I brought him a mirror, the smallest I could find. He looked in it, clutched his head and rocked and rocked on the bed. I went to him and wanted to talk him round. 'When you're a bit better we'll go and live in some abandoned village. We'll buy a house and live there, if you don't want to live in the city with a lot of people. We'll live alone.' I meant it. I would have gone anywhere with him, just so long as he stayed alive, and what he looked like didn't matter. I just wanted him, I meant it.

I won't mention anything I don't want to talk about. There were such things . . . I saw terrible things, perhaps more terrible than death. (*She stops.*)

I was sixteen when we met. He was seven years older than me. We were seeing each other for two years. I really love the area here in Minsk around the main post office, Volodarsky Street. He used to meet me there, under the clock. I lived near the worsted mill and took the No. 5 trolleybus. It didn't stop at the post office, but a bit past it, near the children's clothes store. It slowed down before the turn, which was just what I wanted. I was always a little bit late, so I could look out the window and gasp when I saw how handsome the man waiting for me was! For those two years, I was in a world of my own. I wouldn't have known whether it was winter or summer. He took me to concerts, to see my favourite singer, Édith Piecha. We didn't go running off to dances, because he couldn't dance. We used to kiss, only kiss. He called me 'my little girl'. On my birthday . . . again my birthday . . . It's strange, all the most important things in my life seem to have happened on that day, so I won't let anyone tell me they don't believe in fate. I was standing under the clock. We had a date to meet at five, but he wasn't there. At six, terribly upset, in tears, I wandered back to my bus stop. I was crossing the street, when my sixth sense told me to look round. He was running after me, despite the red light on the crossing, in his boiler suit and boots . . . They wouldn't let him off work any earlier. I really loved the way he looked then. In camouflage fatigues, in a quilted bodywarmer – everything suited him. I went home with him and he changed his clothes. We decided to celebrate my birthday in a restaurant, but couldn't get in. It was already too late in the evening, there were no seats left, and neither of us could afford to slip five or ten roubles (in the old currency) to the man on the door, like other people did. 'Come on!' he said, suddenly beaming. 'Let's buy some champagne and cakes and go to the park. We'll celebrate your birthday there.' Under the stars, under the sky! That's the kind of man he was. We sat on a bench in Gorky Park until morning. I never had another birthday like it in my life. That's when I said to him, 'Marry me. I love you so much!' He laughed and said, 'You're still little.' But the next day, we took our application to the registry office . . .

Oh, how happy I was! I wouldn't change anything in my life, even if someone warned me from above, from the stars, gave me a signal . . . On the day of the wedding, he couldn't find his passport. We turned the whole house upside down looking for it. They had to formalize our marriage at the registry office on some form. 'Oh, daughter, this is a bad omen,' my mother said, in tears. Later, we found the passport in a pair of his old trousers in the attic. Love! Actually, it was more than love: it was a long process of falling more and more in love. How I danced in the morning in front of the mirror: 'I'm pretty, I'm young and he loves me!' Now I'm beginning to forget my face, the face I had when I was with him. I no longer see it in the mirror . . .

Can we talk about this? Put it into words? Some things are secret . . . To this day, I don't really understand what this was. Until our very last month . . . He would call me in the night. He had desires. He loved more strongly even than before. When I looked at him during the day, I couldn't believe what was happening during the nights. We really did not want to be parted from each other. I caressed and stroked him. At those times, I remembered our most joyful, happiest moments. I remembered him coming back from Kamchatka with a beard he had grown there, my birthday on that bench in the park. 'Marry me!' Should I tell you this? Is it all right? I went to him myself, the way a man goes to a woman. What could I give him other than medicines? What hope? He so much didn't want to die. He had a belief that my love would save us. It was such a love! Only I never told my mother anything about it. She wouldn't have understood. She would have condemned me. Cursed me.

This wasn't the normal cancer everybody's afraid of, but Chernobyl cancer, which is even more terrible. The doctors explained it to me: if the metastases had affected him from inside his body, he would have died quickly, but they crawled over the surface. Over his body, over his face. He had a kind of black growth over him. Something happened to his chin, his neck disappeared, his tongue flopped out. His blood vessels burst and he began to haemorrhage. 'Oh dear,' I would cry. 'The bleeding again.' From his neck, his cheeks, his ears. All over the place. I would bring cold water, apply compresses. They didn't really help. It was horrific. The pillow would get soaked. I would bring a basin from the bathroom. The blood dripped into it, like a cow's milk hitting the pail. That sound, so peaceful and rural. I still hear it now, in the nights. While he was conscious he would clap his hands – that was a signal we'd agreed. Phone them! Call an ambulance! He didn't want to die. He was forty-five.

I called the emergency service, but they knew us already and didn't want to come. 'We can do nothing to help your husband.' Well, at least they might give him an injection! Morphine. I would inject him myself. I learned how to do it, but the injection just spread, like a bruise under the skin. It wasn't taken up. One time, I managed to phone through and the ambulance came. It was a young doctor. He approached him, but then backed away. 'Tell me,' he said, 'I don't suppose he's from Chernobyl, is he? Was he one of the people sent there?' I said, 'Yes.' And he literally, I'm not exaggerating, squealed, 'My dear girl, this just needs to end as quickly as possible! The sooner the better! I've seen Chernobyl victims die.' My man was conscious, he heard him say that. The only good thing was that he didn't know, he hadn't guessed, that he was the only member of his team still alive. The last one. Another time, they sent a nurse from the clinic. She stood outside in the hallway and wouldn't even come into the apartment. 'Oh, I can't bear this!' Well, what about me? I could. I could do everything! What could I think of? How could I help him? He was screaming, he was in pain . . . crying out all day. I did find a way: I used the syringe to pour a bottle of vodka into him. It put him out like a light. He could forget everything. I didn't think of that myself, other women told me about it, who had the same problem. His mother would come and say, 'Why did you let him go to Chernobyl? How could you?' But at that time, it never came into my head that there was any possibility of not letting him go – or into his, probably, that he could refuse. It was a different time, like wartime. And we were different then ourselves. I did ask him once, 'Don't you regret now that you went there?' He shook his head. 'No.' He wrote in the notebook, 'When I die, sell the car and the spare wheels, and don't marry Tolik.' That was his brother. Tolik fancied me.

I know secrets . . . I was sitting next to him. He was asleep. He still had lovely hair. I took a pair of scissors and quietly cut off a lock. He opened his eyes, saw what I was holding in my hands, and smiled. What I have left now is his watch, his army record card and his Chernobyl medal . . . (*A long silence.*) Oh, how happy I was! In the maternity hospital, I remember during the day I would sit by the window, waiting for him, looking out. I didn't really understand anything, what was happening or where I was. I just wanted to look at him. I couldn't see enough of him, as if I sensed this must all come to an end soon. In the morning, I would get him his breakfast and gaze at him as he ate it. Watch him shaving, walking down the street. I'm a good librarian, but I can't understand how anyone could passionately love their job. I loved only him. Him alone. And I can't go on without him. I wail at night, scream into the pillow so that the children won't hear . . .

I never for a moment imagined we could be parted, that . . . I knew it, but couldn't picture it. My mother, his brother . . . they were preparing me, hinting that the doctors were saying, advising, counselling . . . In short, there was a special hospital near Minsk where similarly doomed people had been taken in the past . . . veterans of the war in Afghanistan, without arms or legs, and now Chernobyl victims were being sent there. They urged that he would be better off there, with doctors constantly nearby. I didn't want it, wanted to hear nothing about it. Then they worked on persuading him, and he started begging me: 'Send me there. Don't torture yourself.' I was trying to get paid carer's leave, or unpaid leave from work. The law, though, allowed carer's leave only for looking after a sick child, and unpaid leave from work couldn't be for more than a month. He filled up our entire notebook with his pleadings, and eventually made me promise I would take him there. His brother gave me a lift. On the outskirts of a village called Grebyonka stood a big wooden house with a well that had collapsed. There was an outside toilet. Pious old women in black . . . I didn't even get out of the car. I didn't go in. That night, I kissed him and said, 'How could you ask me to agree to that? It's not going to happen! Never!' I kissed him all over.

Those horrible, horrible last weeks . . . Taking half an hour to piss into a half-litre jar. He wouldn't look up, embarrassed. 'How can you think like that?' I kissed him. That last day, there was an amazing moment. He opened his eyes, sat up, smiled and said my name, 'Valyushka!' I was struck dumb with happiness, at hearing his voice again . . .

They telephoned from his workplace: 'We want to bring him a certificate of honour.' I told him, 'Your lads want to come and give you a diploma.' He shook his head. No, and no! But they came anyway . . . brought some money and the certificate, in a red folder with a picture of Lenin. I took it, and thought, 'What is he dying for? In the newspapers, they're writing it's not just Chernobyl but Communism that has blown up. The Soviet way of life is finished, but that profile on the red folder is still the same.' The lads wanted to say some words of appreciation to him, but he hid himself under the blanket, with only his hair showing. They stood over him for a time and then left. He didn't want anyone to see him now. It was only me he would allow to see him. But we die alone. I called him, but he no longer opened his eyes. He was only just breathing . . .

For the funeral, I covered his face with two handkerchiefs. If anyone asked to see him, I drew them back. One woman collapsed . . . She had once loved him and I was jealous of her. 'Let me see him one last time.' 'Go ahead.'

I didn't tell you that, when he died, no one could bring themselves to go near him. They were all afraid. And family members can't lay out a body themselves. It's our Slavic tradition. They brought two male nurses from the mortuary. They asked for vodka: 'We thought we'd seen it all,' they admitted, 'mangled corpses, knifed, bodies of children after a fire, but we've never seen anything like this . . .' (*She quietens down.*) He died and lay there, so hot you couldn't touch him. I stopped all the clocks in the house . . . It was seven in the morning . . . our clocks are all stopped to this day, you can't restart them . . . The clock repairer came. He just shrugged and said, 'It's not mechanical, not physics. It's metaphysics.'

Those first days, without him . . . I slept for two days. No one could wake me. I would get up, have a drink of water, not even eat, and then fall back on the pillow again. It seems strange to me now that I could sleep like that. When my friend's husband was dying, he threw plates at her. He cried. He wanted to know why she was so young and beautiful. My husband just gazed and gazed at me. He wrote in our notebook, 'When I die, have my remains cremated. I don't want you to be frightened.' Why did he make that decision? Well, there were various rumours around: even after death, Chernobyl victims were said to glow . . . At night, a light would appear above their graves. I had read myself that people gave a wide berth to the graves of the Chernobyl firemen who had died in Moscow hospitals and been buried nearby in Mitino. Local people wouldn't bury their own dead alongside them. The dead afraid of the dead . . . to say nothing of the living. Because nobody yet understands Chernobyl. It's all speculation, dread. He brought his white work clothes back from Chernobyl. Trousers, boiler suit. They were in our top cupboard until he died. Then my mother decided, 'We need to throw out all his belongings.' She was frightened, but I kept even his boiler suit. 'You're a criminal! You have young children in the house. A son and daughter.' We took them outside the city and buried them. I've read a lot of books, I live among them, but they don't explain anything. They brought us the urn. I wasn't afraid of it. I touched the ashes and found something small in there, like seashells on the beach, in the sand. It was fragments of hip bone. Until then, when I touched his things, I hadn't felt, hadn't sensed anything, but then it was as if I had put my arms round him. During the night, I remember, he was dead and I was sitting beside him. Suddenly, there was a haze. I saw it above him a second time at the crematorium . . . His soul . . . No one else saw it, only me. I felt we had seen each other once more . . .

Oh, how happy I was, how happy! If he had to go away for his work, I counted the days and hours, the seconds, until we would be together again! I physically can't do without him, I just can't! (*Covers her face with her hands.*) I remember we went to visit his sister in the countryside. In the evening, she said, 'I've made up this room for you, and that room for him.' We looked at each other and burst out laughing. It had never occurred to us that we could sleep apart, in different rooms. We had to be together. I can't live without him. I can't! I've had plenty of offers, including from his brother. They're so similar, in height, even the way they walk. But it seems to me that, if anyone else were to touch me, I would just cry and cry and never stop.

Who took him away from me? What right did they have? They handed him a call-up notice with a red stripe, dated 19 October 1986 . . . (*She brings a photograph album and shows wedding pictures. When I want to say goodbye, she stops me.*)

How can I go on living? I haven't told you everything, not all of it. I was happy, madly happy. There are secrets . . . perhaps you shouldn't include my name . . . People say their prayers in private. To themselves . . . (*Trails off.*) No, put my name! It will be a reminder to God . . . I want to know, I want to understand why we should have to bear this sort of suffering? What's it for? At first, it seemed that after everything that had happened something dark would appear in my eyes, something alien. I wouldn't be able to endure it. What saved me? What forced me back towards life? Brought me back? My son. I have another son . . . The first boy I had with him. He's been ill for a long time. He's grown up now, but sees the world through the eyes of a child, the eyes of a five-year-old. I want to be with him now. I hope to exchange my apartment for one closer to Novinki, the mental hospital there, where he's lived all his life. That was the verdict of the doctors: for him to live, he needed to be there. I go every day. When he sees me, he asks, 'Where is Daddy Misha? When will he come?' Who else could ask me that? He's waiting for him.

We will wait for him together. I will say my Chernobyl prayer, and he will look at the world with the eyes of a child . . .

Valentina Timofeyevna Apanasevich, wife of a clean-up worker

In place of an epilogue

From materials published in Belarusian newspapers in 2005

. . . Kiev travel agency offers tourist trips to Chernobyl . . .

We have arranged an itinerary, starting in the ghost city of Pripyat. You will visit abandoned multi-storey apartment blocks with blackened laundry on their balconies and children's prams. The former police station, hospital and the Municipal Communist Party Committee building. The slogans of the Communist era are still there: not even radiation can obliterate them.

From Pripyat, our route takes us to ghost villages where wolves and wild boar scavenge through the cottages in broad daylight. They have been proliferating, monstrously!

The culmination of the trip or, as they say in the brochure, the 'highlight', is a viewing of the Shelter Reactor, more commonly known as the 'Sarcophagus'. Hastily constructed over the destroyed Reactor No. 4, it has long been covered in cracks through which its deadly contents emit background radiation from what remains of the nuclear fuel. You are certainly going to have something to tell your friends about when you get back home. This is not just some excursion to the Canary Islands or Miami . . .

Your trip concludes with souvenir photographs at the stele in memory of the fallen heroes of Chernobyl, to give you a sense of your involvement in history.

After the excursion, lovers of extreme tourism are invited to enjoy a picnic lunch of ecologically safe food, washed down with red wine. And Russian vodka. You are assured that, during your day spent in the exclusion zone, you will be subjected to a dose of radiation less than that involved in a standard medical X-ray check-up. You are, however, advised not to bathe, or eat any fish you have caught or game you have shot. Neither should you pick berries or mushrooms, or roast them on a campfire. You should also not give ladies bunches of wild flowers.

Do you think I am out of my mind? You would be mistaken: atomic tourism is in great demand, especially among Westerners. People crave strong new sensations, and these are in short supply in a world so much explored and readily accessible. Life gets boring, and people want a frisson of something eternal . . .

Visit the atomic Mecca. Affordable prices.